# The Binary Foundation of Intelligence

*From Bits and Bytes to Artificial Intelligence*

**By Susie Hala**

**IntelliGloss Press**

United States of America

# Disclaimer

**Educational Purpose Disclaimer**
"This publication is part of the IntelliGloss AI Education Series, a structured curriculum support system for AI literacy in grades 6–12."

The information presented in this publication is provided for educational and informational purposes only. This book is intended to support educators, administrators, parents, and learners in understanding the concepts, opportunities, and challenges associated with artificial intelligence in education. It is not intended to serve as legal, technical, policy, cybersecurity, or professional advice.

Readers should consult qualified professionals, institutional guidelines, and local regulations before implementing any educational programs, technologies, or policies described in this book.

**Technology and Accuracy Disclaimer**
Artificial intelligence is a rapidly evolving field. While every effort has been made to ensure the accuracy and reliability of the information presented at the time of publication, technologies, policies, research findings, and best practices may change over time. The author and publisher make no guarantees that all information will remain current or applicable in future technological environments.

Readers are encouraged to verify tools, platforms, and recommendations independently before using them in educational settings.

**Implementation Responsibility**
Schools, educators, and institutions are solely responsible for determining how and whether to implement artificial intelligence tools, strategies, or classroom activities discussed in this book. Educational environments vary widely, and policies regarding technology use, student privacy, and curriculum development differ by district, state, and country.

The author and publisher assume no responsibility for outcomes resulting from the implementation of ideas, examples, or frameworks described in this book.

**Student Safety and Data Privacy**
Artificial intelligence tools may involve the collection or processing of data. Educators and institutions are responsible for ensuring that any technology used complies with applicable student privacy laws, data protection regulations, and school policies, including but not limited to FERPA, COPPA, GDPR, or other relevant regulations depending on jurisdiction.

The author and publisher are not responsible for the privacy practices, security policies, or data handling procedures of third-party platforms or software mentioned in this book.

**Third-Party Tools and References**
This book may reference third-party technologies, organizations, research studies, or platforms for illustrative and educational purposes. Such references do not constitute endorsements or guarantees regarding the performance, reliability, safety, or suitability of those tools for educational use.

Educators and institutions should conduct their own evaluations and risk assessments before adopting any technology.

**No Professional Liability**
The author and publisher shall not be held liable for any direct, indirect, incidental, or consequential damages resulting from the use or misuse of the information contained in this book. This includes but is not limited to decisions related to curriculum design, technology adoption, classroom instruction, institutional policies, or educational outcomes.

**Educational Interpretation**
The concepts and examples provided in this book are intended to support discussion and learning. They should not be interpreted as official standards, mandatory guidelines, or universally applicable policies for artificial intelligence education.

Schools and educators are encouraged to adapt the ideas presented here according to their local policies, cultural context, educational standards, and student needs.

**Author's Intent**
The goal of this book is to promote responsible AI literacy, thoughtful discussion, and ethical awareness as schools prepare students for a world increasingly shaped by intelligent technologies.

The author encourages readers to approach artificial intelligence education with curiosity, caution, and a commitment to responsible learning.



**Title:** *The Binary Foundation of Intelligence: From Bits and Bytes to Artificial Intelligence*
**ISBN:** 978-1-972925-00-3

Published by
IntelliGloss Press
United States of America

Cover design and layout by IntelliGloss Press
First Edition
2026

Printed in the United States of America



## Dedication

This book is dedicated to the next generation of learners—
the students who will grow up in a world shaped by intelligent technology,
and who deserve to understand it with clarity, confidence, and purpose.

To the educators who show up every day with dedication and heart,
guiding students through a rapidly changing world—
your role has never been more important.

To the parents and families who support learning at home,
encouraging curiosity, responsibility, and growth—
you are the foundation of every child's success.

And to my granddaughters,
whose curiosity, imagination, and future inspire me to do this work—
this book is for you.

May you not only learn how to use technology,
but understand it, question it, and shape it for the good of others.

## About the Author

**Susie Hala**
Founder of the IntelliGloss AI Education Series

Susie Hala is an author, educator, and curriculum developer dedicated to advancing artificial intelligence literacy in education. Her work focuses on helping schools, educators, and students understand how intelligent technologies are shaping modern society—and why that understanding is essential for the next generation.

Hala is the creator of the IntelliGloss AI Education Series, a comprehensive and growing collection of educational books designed to make complex artificial intelligence concepts accessible, structured, and practical for classroom use. Her work emphasizes clarity, organization, and responsible implementation, enabling schools to introduce AI literacy without requiring advanced technical backgrounds.

In addition to her work in education, Hala brings hands-on experience working with artificial intelligence systems, including early exposure to rule-based AI technologies during her time working with AT&T. She has also developed modern digital platforms that integrate artificial intelligence, automation, and online systems, building e-commerce solutions and AI-driven applications that connect payment platforms, hosting environments, and conversational AI tools into real-world, functional systems.

This combination of foundational understanding and practical experience allows her to translate complex AI concepts into clear, accessible language for both educators and students.

Recognizing that artificial intelligence is rapidly transforming how information is created, interpreted, and used, Hala advocates for equipping students with the ability not only to use AI tools, but to understand, question, and evaluate them. Her work emphasizes critical thinking, ethical awareness, and responsible digital citizenship.

The IntelliGloss AI Education Series supports schools in teaching key areas such as artificial intelligence foundations, algorithms, machine learning concepts, and AI ethics, as well as the broader evolution of intelligent systems—from Artificial Narrow Intelligence (ANI) to Artificial General Intelligence (AGI) and Artificial Superintelligence (ASI). These resources are designed for students in grades 6–12 and for educators seeking structured, classroom-ready approaches to AI literacy.

In addition to her work in education publishing, Hala has a strong background in entrepreneurship, having built and managed multiple businesses and educational initiatives. Her experience across industries reinforces her belief that technology education must remain grounded in human values, critical thinking, and ethical responsibility.

**Her mission is clear: to ensure that the next generation understands intelligent technology—not just how to use it, but how it truly works.**

## Preface

*Two Students. Two Paths. One Future.*

Before we begin, it is important to understand why learning how artificial intelligence works—not just how to use it—matters for every student.

Imagine two students who have just graduated from high school and are preparing for a job interview. The first student knows how to use artificial intelligence tools. This student can generate answers, complete assignments quickly, and rely on AI to assist with tasks. However, this student does not understand how artificial intelligence works behind the scenes.

The second student also knows how to use AI tools, but in addition, understands the foundational concepts behind them. This student has learned about data, algorithms, tokens, models, and how intelligent systems are trained and operate. This student can think critically about AI, recognize its limitations, and adapt when needed.

At first, the first student may appear faster and more efficient. However, over time, the second student demonstrates a deeper level of understanding, independence, and problem-solving ability. The second student is not only able to use artificial intelligence but also to question it, improve it, and grow with it.

This book is written for the second student.

In a world where artificial intelligence is becoming part of everyday life, students must not only learn how to use intelligent systems—they must learn how they work. True understanding leads to confidence, responsibility, and the ability to lead in an AI-driven future.

This perspective is not just theoretical—it is personal.

There was a time when I was the first student.

I knew how to use systems. I followed processes, completed tasks, and worked within structured environments that required accuracy and consistency. At the time, I did not think of these systems as artificial intelligence. They were simply part of the job—tools that had to be learned, followed, and mastered.

But looking back, I now understand something much deeper. I was working within rule-based systems—the earliest foundation of what we now recognize as artificial intelligence.

That realization changed everything.

As I began studying modern AI, including machine learning, deep learning, and intelligent systems, I started to see a clear connection between the past and the present. The structured systems I once worked in, the logic behind decision-making, and the step-by-step processes all still exist today. They have simply evolved.

Artificial intelligence did not appear overnight. It grew from rule-based systems into systems that can now learn, adapt, and assist in ways we never imagined.

I made a decision to move from simply using systems to understanding them.

That journey—from student one to student two—is the reason these books exist.

My mission is to help others bridge that same gap.

I believe that artificial intelligence should be understandable for everyone. It should not feel intimidating or out of reach. Students should feel confident exploring it. Teachers should feel prepared to introduce it. Everyday learners should have access to clear, structured explanations that make sense.

You do not need a background in technology. You do not need to know how to code. You only need the right explanation.

**The IntelliGloss Instructional Approach**

The IntelliGloss AI Education Series is designed to help students move beyond simply using AI tools and begin understanding artificial intelligence as a complete system.

Students learn how AI is built, how it operates, and how it connects to the real world.

## A Message to Students and Readers

If you have ever felt that technology was too complicated…
If you have ever been confused by systems that were never clearly explained…

I want you to know:

You are capable of understanding artificial intelligence.

Sometimes, the challenge is not the concept itself—it is how it has been presented.

This series was created to change that.

## A Message to Educators

To the educators using this material:

These books are designed to support you in introducing artificial intelligence in a way that is clear, structured, and meaningful for students.

AI is becoming part of every industry, and students deserve more than just exposure to tools—they deserve understanding.

This series aims to provide that foundation.

## Final Reflection

Every experience we have—especially the challenging ones—prepares us for something greater.

The systems I once worked in taught me discipline, structure, and attention to detail. Today, those same lessons allow me to break down complex ideas and make them understandable for others.

That is what this series represents:

Turning complexity into clarity.

## *Why Artificial Intelligence Education Matters for Every Student*

### Why AI Literacy Is Essential for Every Student

Artificial intelligence is no longer a future concept or a specialized technology used only by experts. It is already part of everyday life for students of all ages. From smartphones and search engines to navigation apps, social media, streaming platforms, and educational tools, AI systems influence how information is accessed, decisions are made, and learning takes place.

Because AI is embedded in daily life, understanding it is no longer optional. It is a foundational skill, similar to digital literacy and critical thinking. Students do not need to become programmers or engineers, but they do need to understand what AI is, how it works at a basic level, and how it affects their lives.

AI education should not be limited to computer science classrooms. Every student interacts with AI systems daily. Teaching students how these systems function empowers them to become informed users rather than passive consumers of technology.

This book is designed to make artificial intelligence understandable, approachable, and relevant. It focuses on clarity, responsibility, and real-world understanding to help students prepare for a future where humans and intelligent systems work side by side.

## IntelliGloss AI Education Series

**Teaching how AI actually works—from hardware to power to global systems.**

## Why AI Education Matters

Artificial intelligence is already shaping how students learn, write, research, and solve problems.

Yet many students—and educators—are using AI tools without understanding how these systems actually work.

Schools need more than access to technology.
They need structured, responsible AI education.

## Our Approach

The IntelliGloss AI Education Series is designed to help students move beyond simply using AI tools and begin understanding artificial intelligence as a complete system.

Students learn how AI is built, how it operates, and how it connects to the real world.

## The IntelliGloss Framework

Our books introduce AI through a clear and connected system:

**Chip Family** — the hardware that powers AI

**Watts Family** — the rate at which AI uses power

**Energy Family** — the total power AI consumes over time

**Infrastructure Family** — the global systems that support AI

This framework helps students understand AI from the inside out.

## Built for the Classroom

Each book includes:

Structured chapters aligned with semester learning

Clear explanations designed for student understanding

Reflection sections with learning objectives

Worksheets, quizzes, and answer keys

Teacher Implementation Guides

### Supporting Schools and Educators

The IntelliGloss series supports:

- Responsible and ethical AI use
- Student data awareness and safety
- Critical thinking and digital literacy
- Classroom-ready instruction for grades 6–12 and beyond

### Our Goal

To help schools transition from simply using AI tools to truly understanding artificial intelligence as a complete system.

### Closing Statement

IntelliGloss is more than a book series.
It is a structured approach to preparing students for a future shaped by artificial intelligence.

In today's world, many programs focus on teaching how to use artificial intelligence tools. IntelliGloss takes a different approach by helping students understand how AI actually works—from its foundational building blocks to real-world applications.

## Who This Book Is For

This book is designed for educators, school leaders, and curriculum developers guiding students in an age of rapidly advancing technology.

Artificial intelligence is already influencing how students learn, research, and communicate. Yet many educators have not received formal training in how these systems work or how they should be addressed in the classroom.

This book helps bridge that gap by providing clear explanations and practical frameworks for understanding artificial intelligence in an educational setting.

### How Schools Can Use This Book

This book supports schools in developing a structured approach to artificial intelligence literacy.

It can be used for:

Professional development for educators

Classroom discussions and activities

Curriculum support across subject areas

Policy and academic integrity conversations

Rather than focusing on specific tools, this book emphasizes lasting principles that help students understand, evaluate, and use AI responsibly.

## Closing Perspective

Artificial intelligence will continue to evolve. With the right knowledge and leadership, schools can ensure that students not only use these technologies, but understand them, question them, and apply them responsibly.

IntelliGloss is more than a book series. It is a structured approach to preparing students for a future shaped by artificial intelligence.

Rather than focusing only on how to use AI tools, this series helps students understand how AI actually works—from its foundational building blocks to real-world applications.

## Human Values Family

## Definition

The Human Values Family in artificial intelligence represents the guiding principles that shape how AI systems are created, used, and integrated into society. These values include ethics, responsibility, trust, and human well-being.

Artificial intelligence may be built on data, algorithms, and computational systems, but its true impact is measured by how it affects people. For this reason, AI is not only a technical system—it is a human-centered system.

## Key Concepts

The Human Values Family exists to ensure that artificial intelligence serves humanity in a positive and responsible way. While AI systems can process information and generate outcomes, they do not possess judgment, empathy, or moral understanding.

This means that human values must guide every stage of AI development—from design to deployment.

AI reflects the intentions, decisions, and assumptions of the people who build and use it. If those values are not carefully considered, AI systems can unintentionally produce biased, harmful, or misleading results.

Understanding human values in AI is not optional—it is essential.

## Core Elements of the Human Values Family

**Ethics**
Ethics in AI refers to making decisions that are fair, just, and respectful of human rights. This includes minimizing bias, protecting privacy, and ensuring transparency in how systems operate.

**Responsibility**
Responsibility means that humans remain accountable for AI systems. Developers, educators, organizations, and users must take ownership of how AI is used and the outcomes it produces.

**Trust**
Trust is critical for the successful adoption of AI. People must feel confident that AI systems are reliable, safe, and aligned with their best interests.

**Human Well-Being**
AI should enhance human life—not replace, harm, or diminish it. This includes supporting learning, improving access to information, and creating opportunities while maintaining human dignity.

## Real-World Connection

Artificial intelligence is already influencing how people learn, communicate, and make decisions. From search engines to classroom tools, AI is shaping daily experiences.

Without clear human values guiding these systems, there is a risk of reinforcing misinformation, bias, or unrealistic expectations.

For students, this means learning not only how to use AI tools, but how to question them, evaluate their outputs, and understand their limitations.

## Applications in Education

In the classroom, the Human Values Family helps students develop critical thinking and ethical awareness. Teachers can use this framework to guide discussions about fairness, responsibility, and the impact of technology on society.

Students can explore questions such as:

How should AI be used responsibly?

What makes an AI system fair?

How does technology influence human behavior?

This approach supports both technical understanding and responsible decision-making.

## Benefits

Teaching human values in AI helps students:

Develop ethical awareness

Strengthen critical thinking

Understand the societal impact of technology

Become responsible users and future creators of AI

It ensures that learning about AI goes beyond functionality and includes responsibility.

## Challenges

Human values are not always universal. Different cultures, communities, and individuals may have different perspectives on what is considered fair or ethical.

Additionally, AI technology is evolving rapidly, often faster than policies and guidelines can keep up. This makes it essential for educators and students to stay informed and adaptable.

## Summary

The Human Values Family reminds us that artificial intelligence is not just about machines—it is about people.

By focusing on ethics, responsibility, trust, and human well-being, students can learn to use AI thoughtfully and responsibly.

Understanding these principles helps prepare learners to navigate a world where artificial intelligence plays an increasingly important role in everyday life.

# TABLE OF CONTENTS

## Connecting AI Understanding to Cybersecurity Awareness

### Definition

Understanding artificial intelligence is the first step toward understanding cybersecurity. While cybersecurity focuses on protecting systems, data, and users, artificial intelligence explains how those systems are built, how they function, and where they may be vulnerable.

### The Connection

Every AI system is made up of components such as data, algorithms, models, and infrastructure. These same components are also the targets of cybersecurity threats.

When students understand:

How data is collected and stored

How algorithms make decisions

How AI models process information

How systems operate across frontend and backend layers

They begin to recognize where risks can exist and how systems can be misused.

### Why This Matters

In today's digital world, students are not just users of technology—they are participants in complex systems powered by artificial intelligence.

Without understanding how these systems work:

Technology becomes something they trust without question

Risks become harder to recognize

Decisions are made without awareness of consequences

With understanding:

Students become more aware of how their data is used

They recognize how systems can be manipulated

They make more informed and responsible choices

### Building Digital Readiness

Digital readiness is not only about using tools. It is about understanding the systems behind those tools.

By learning how AI works, students develop:

Awareness of digital environments

Confidence in navigating technology

A foundation for future learning, including cybersecurity

### Conclusion

Cybersecurity begins with understanding.

Before students can protect systems, they must first understand how those systems are built and how they operate. This is why learning artificial intelligence is not only about innovation—it is also about responsibility.

## The Foundational Families of Artificial Intelligence

### Introduction

Artificial intelligence is often described as a powerful technology, but in reality, it is not a single system. It is a combination of multiple components working together to create intelligent behavior.

To better understand how artificial intelligence works, it is helpful to break it down into organized groups. In the IntelliGloss framework, these groups are called **Foundational Families of Artificial Intelligence**.

Each family represents a key part of how AI systems are built and operate. Some families focus on how information is created and stored. Others focus on how machines learn, make decisions, process language, or perform tasks. Additional families represent the physical hardware, energy systems, and global infrastructure that make artificial intelligence possible.

When these families work together, they form a complete AI system.

This approach allows students and educators to move beyond simply using artificial intelligence tools and begin to understand what is happening behind the scenes. Instead of viewing AI as a mystery, learners can see it as a structured system made up of clearly defined components.

The diagram on the next page provides a visual overview of these families and how they connect to form the foundation of artificial intelligence.

By learning each family step by step, students develop a deeper understanding of how AI systems process information, learn from data, and interact with the world.

Understanding the Foundational Families of Artificial Intelligence is the first step toward true AI literacy.

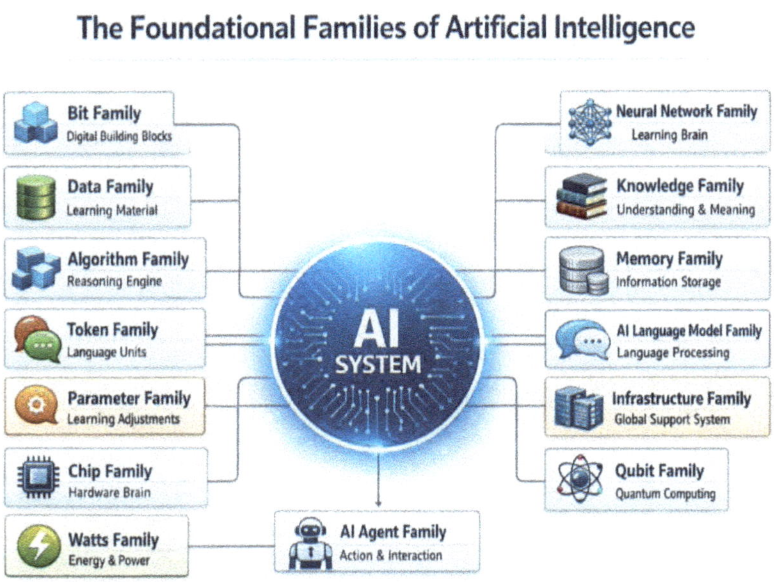


## The Foundational Families of Artificial Intelligence

| Family | Role | Description |
|---|---|---|
| **Bit Family** | Digital Building Blocks | The smallest unit of digital information (0 or 1). All computing systems begin with bits. |
| **Algorithm Family** | Reasoning Engine | Step-by-step instructions that tell computers how to solve problems and process information. |
| **Token Family** | Language Units | Pieces of text used by AI models to understand and generate human language. |

| Family | Role | Description |
|---|---|---|
| **Parameter Family** | Learning Adjustment System | Adjustable numerical values inside AI models that allow them to learn patterns from data. |
| **Data Family** | Learning Material | The information AI studies in order to learn patterns, relationships, and knowledge. |
| **Logic Family** | Decision Rules | The reasoning structures that allow AI systems to make decisions and evaluate conditions. |
| **Memory Family** | Information Storage | Systems that store and retrieve information so AI models can access past knowledge and context. |
| **Knowledge Family** | Understanding & Meaning | Structured information that allows AI to represent facts, relationships, and concepts. |
| **Neural Network Family** | Learning Brain | Interconnected layers of artificial neurons that enable AI systems to recognize patterns, learn from data, and make predictions. |
| **AI Model Family** | Intelligence Engine | The trained system that uses data, parameters, and neural networks to perform tasks such as prediction, classification, and generation. |
| **AI Translator Architecture Family** | Language Conversion System | Systems that transform input into output, enabling AI to convert text, speech, or images into meaningful responses. |
| **AI Agent Family** | Action and Task Execution System | AI systems designed to perceive information, make decisions, and carry out tasks autonomously or semi-autonomously. |
| **Chip Family** | Hardware Brain | Physical processors (GPUs, TPUs, CPUs, NPUs) that perform AI computations. |
| **Watts Family** | Energy Power | The electrical power required to run AI hardware and data centers. |
| **AI Energy Family** | Energy Source System | The total energy consumed over time to operate AI systems, including training and real-time usage. |
| **Infrastructure Family** | Global Support System | Data centers, internet networks, cloud systems, and global infrastructure that allow AI to operate and scale. |
| **Ingredient Family** | System Composition | The essential components that come together to create an AI system, including data, algorithms, models, hardware, and energy. |
| **Qubit Family** | Quantum Information Units | Quantum bits that can represent multiple states simultaneously, forming the foundation of quantum computing. |

**To begin understanding how artificial intelligence is built, we start with the most fundamental unit of all computing systems: the bit.**

## The Bit Family

## Introduction

The Bit Family represents the starting point of all digital technology. Every computer system, application, and artificial intelligence model is built from bits—the smallest unit of information in computing.

A bit can hold one of two values: 0 or 1. While this may seem simple, these two values form the foundation of everything digital. When bits are combined and organized, they create larger units of data that allow computers to store text, display images, play videos, and run complex systems.

From a single bit to massive data systems, all digital information follows this same structure. This is how computers are able to represent and process the world in a form they can understand.

In artificial intelligence, bits are essential because they store the data that models learn from, process the calculations that drive decision-making, and support the systems that generate responses.

Understanding the Bit Family helps students see how simple binary signals grow into powerful technologies. It reveals that behind every advanced AI system is a foundation built from the most basic building blocks of information.

# Bit Family — Visual Understanding of Data Size

## How Much Data Can Each Unit Hold? (Simple Examples)

| Unit | Size | Photos (Approx.) | Video (Approx. Hours) |
|---|---|---|---|
| Bit | 1 bit | N/A | N/A |
| Byte | 8 bits | 1 character | N/A |
| Kilobyte (KB) | ~1,000 bytes | Part of a paragraph | N/A |
| Megabyte (MB) | ~1,000 KB | ~1 photo | ~1 minute of video |
| Gigabyte (GB) | ~1,000 MB | ~250 photos | ~1–2 hours of video |
| Terabyte (TB) | ~1,000 GB | ~250,000 photos | ~1,000–2,000 hours of video |
| Petabyte (PB) | ~1,000 TB | ~250 million photos | ~1–2 million hours of video |
| Exabyte (EB) | ~1,000 PB | ~250 billion photos | ~1–2 billion hours of video |
| Zettabyte (ZB) | ~1,000 EB | ~250 trillion photos | ~1–2 trillion hours of video |
| Yottabyte (YB) | ~1,000 ZB | ~250 quadrillion photos | ~1–2 quadrillion hours of video |

---

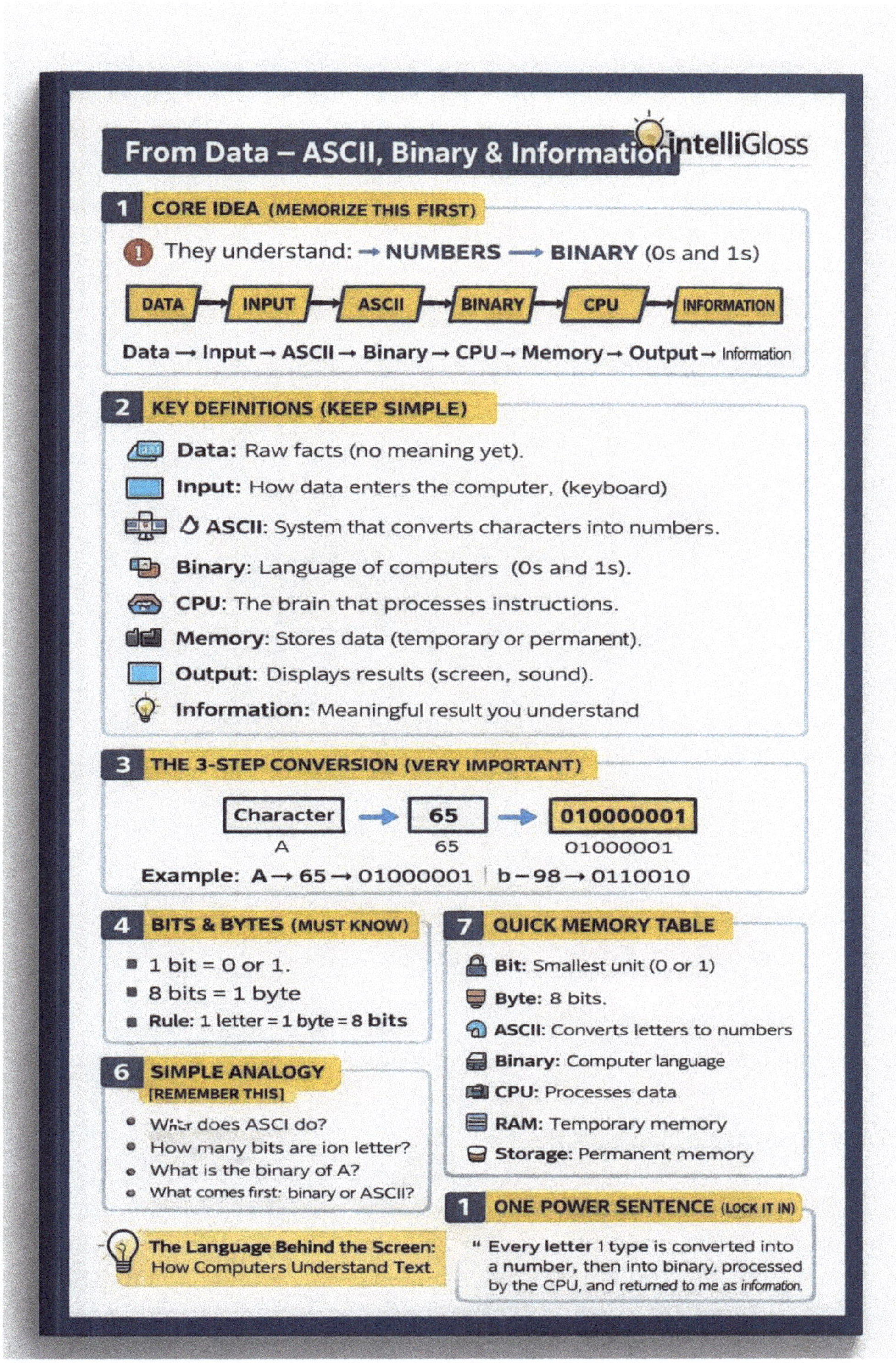


## ⚙ *The Algorithm Family*

## Definition and Overview

The **Algorithm Family** represents the *thinking engine* of artificial intelligence — the set of mathematical instructions that guide how AI analyzes data, learns patterns, and makes decisions.

An **algorithm** is a sequence of steps designed to solve a problem or complete a task. In AI, algorithms tell the system *how to learn*, *how to predict*, and *how to improve* with every iteration.

In simple terms:

The Algorithm Family is the *logic of intelligence* — the invisible code that transforms data into decisions and predictions into progress.

## 🧠 How Algorithms Power AI

Every AI system — from simple chatbots to large-scale deep learning models — relies on algorithms as its foundation.
They determine how the model:

- Processes and cleans data,
- Identifies relationships between variables,
- Learns from feedback, and
- Improves over time.

Without algorithms, even the most powerful dataset or neural network would remain static and lifeless.

In essence, **algorithms are the teachers** that train AI how to think.

## ⚙️ *Why the Algorithm Family Matters*

Algorithms are what separate automation from intelligence.
They allow machines to **reason**, **adapt**, and **learn** — not just follow commands.
For prompt engineers, understanding algorithmic behavior helps explain *why* AI sometimes misunderstands a question, repeats patterns, or changes tone after feedback.

Every AI response you receive follows an underlying process — a pattern recognition journey from input to output.

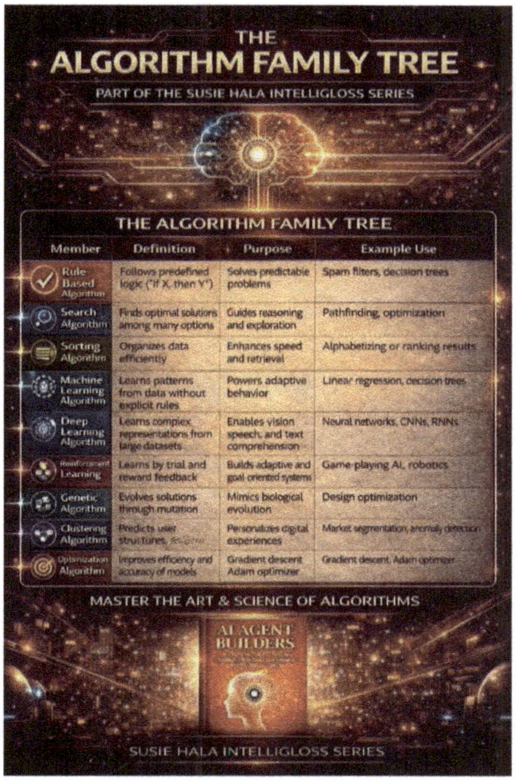


## 🌍 Analogy for Everyday Readers

Imagine AI as a chef in a kitchen:

- The **Dataset Family** provides the ingredients.
- The **Knowledge Family** gives the recipes.
- The **Algorithm Family** is the *cooking process* — mixing, heating, tasting, and adjusting until the dish is perfect.

Just as cooking methods determine the quality of food, algorithms determine the quality of intelligence.

## ⚡ Algorithm Efficiency and Energy Use

Not all algorithms are equal.
Some are **fast but shallow**, while others are **deep but energy-intensive**.
Training large AI models requires algorithms that can handle billions of calculations per second — powered by the **Watts Family** (energy) and executed through the **Chip Family** (hardware).

The design of efficient algorithms directly affects sustainability in AI — minimizing power consumption while maximizing accuracy and speed.

### 🧩 Algorithms and Prompt Engineering

For prompt engineers, algorithms explain *how* the AI arrives at its answers.
When a model gives inconsistent or off-topic results, it's often due to:

Misinterpretation of input signals (input encoding),

Bias in training algorithms, or

Overfitting to certain data patterns.

Knowing this helps you craft prompts that guide the AI's decision-making — not just its output.

💬 **Quick Takeaway: The Algorithm Family is the reasoning heart of AI.**
It transforms information into understanding and enables machines to think, learn, and evolve like digital problem-solvers.

Absolutely, partner — this is a powerful section in your book because **tokens are where AI meets language**. I'll keep this aligned with your style: clear, structured, and school-district ready.

## Token Family Introduction

The Token Family represents one of the most important bridges between human language and artificial intelligence. While humans communicate using words, sentences, and meaning, AI systems do not understand language in the same way. Instead, they rely on tokens—small units of text that allow machines to process, analyze, and generate language step by step.

A token can be a word, part of a word, a character, or even punctuation. When a student types a sentence into an AI system, that sentence is not read as a whole idea. It is first broken down into tokens. These tokens become the input that the AI model uses to recognize patterns, predict outcomes, and generate responses.

The Token Family works closely with several other foundational families in artificial intelligence. It connects directly with the Data Family, which provides the text used for training, and the Neural Network Family, which processes token patterns to produce intelligent outputs. It also plays a key role in the AI Translator Architecture Family, where tokens are transformed, analyzed, and reassembled into meaningful responses.

Understanding tokens helps students see what is happening behind the scenes when they interact with AI tools. Instead of viewing AI as a "magic system," they begin to understand that every response is built from sequences of tokens processed through mathematical models.

In simple terms, if language is what humans speak, tokens are what AI understands.

### Token Family Breakdown Chart

| Component | Role | Description | Example |
|---|---|---|---|
| Tokenization | Input Conversion | The process of breaking text into smaller units (tokens) that AI can process | "Artificial Intelligence" → "Artificial" + "Intelligence" |
| Tokens | Language Units | The individual pieces of text used by AI models for understanding and generation | "AI", "learn", "ing", "." |
| Subword Tokens | Efficiency Units | Words split into smaller parts to handle unknown or complex vocabulary | "unbelievable" → "un", "believ", "able" |
| Character Tokens | Fine-Grained Units | Text broken down into individual characters for detailed processing | "AI" → "A" + "I" |
| Vocabulary (Token Set) | Token Library | The complete set of tokens an AI model recognizes and uses | GPT models have thousands to millions of tokens |
| Encoding | Token Mapping | Converting tokens into numerical representations for computation | "AI" → [1234, 5678] (example IDs) |
| Decoding | Output Reconstruction | Converting numerical outputs back into readable text | [1234, 5678] → "AI system" |
| Context Window | Memory Limit | The number of tokens an AI can process at one time | Example: 8,000 tokens ≈ several pages of text |
| Token Embeddings | Meaning Representation | Mathematical vectors that capture relationships between tokens | "king" and "queen" have similar embeddings |
| Token Prediction | Language Generation | The process of predicting the next token in a sequence | "AI is" → predicts "powerful" |

The Token Family reminds us that behind every intelligent response is a structured sequence of tokens working together. What feels like natural conversation to humans is, for AI, a carefully processed stream of language units transformed into meaning.

## Parameter Family Introduction

The Parameter Family represents the internal learning system of artificial intelligence. While tokens provide the language and data provides the learning material, parameters are what allow AI systems to adjust, improve, and make intelligent decisions over time.

In simple terms, parameters are numerical values inside an AI model that change as the system learns. These values are not visible to users, but they play a critical role in how the model understands patterns, relationships, and meaning. Every time an AI system is trained, it adjusts its parameters to better predict outcomes and generate accurate responses.

Parameters exist throughout neural networks, where they act as the connections between artificial neurons. Each connection has a weight, which determines how important a particular piece of information is. During training, these weights are continuously adjusted based on data, allowing the system to improve its performance.

The Parameter Family works closely with the Data Family, which provides the information needed for learning, and the Neural Network Family, which organizes how parameters are structured and applied. It also supports the Token Family by helping the model interpret and predict sequences of language.

Understanding parameters helps students recognize that AI is not simply programmed with fixed rules. Instead, it learns by adjusting internal values that guide how it processes information. These adjustments are what allow AI systems to evolve from simple pattern recognition to complex reasoning and decision-making.

In essence, if data is what AI learns from, parameters are how AI learns.

## Parameter Family Breakdown Chart

| Component | Role | Description | Example |
|---|---|---|---|
| Parameters | Learning Values | Numerical values inside an AI model that are adjusted during training | Millions or billions of values inside a model |
| Weights | Connection Strength | Values that determine how strongly one neuron influences another | A higher weight = stronger influence |
| Bias | Adjustment Factor | A value added to help the model shift and fine-tune outputs | Helps improve accuracy in predictions |
| Training Process | Learning Mechanism | The process of adjusting parameters using data to improve performance | Model improves after analyzing many examples |
| Loss Function | Error Measurement | Measures how far the model's prediction is from the correct answer | Lower loss = better performance |
| Optimization Algorithm | Adjustment Strategy | Method used to update parameters efficiently | Gradient Descent adjusts weights step by step |
| Gradient | Direction of Change | Indicates how parameters should change to reduce error | Shows whether to increase or decrease a value |
| Backpropagation | Learning Process | System that sends error signals backward to update parameters | Adjusts earlier layers based on output error |

| Component | Role | Description | Example |
|---|---|---|---|
| Hyperparameters | Control Settings | External settings that guide how training happens (not learned) | Learning rate, batch size |
| Model Size | Capacity Indicator | The total number of parameters in a model | Larger models = more learning capacity |

The Parameter Family is the hidden engine of learning in artificial intelligence. While users interact with outputs, it is the continuous adjustment of parameters behind the scenes that makes intelligent behavior possible.

## Data Family Introduction

The Data Family represents the foundation of learning in artificial intelligence. Every AI system depends on data to develop knowledge, recognize patterns, and make decisions. Without data, artificial intelligence cannot learn, adapt, or function effectively.

Data can take many forms, including text, images, audio, video, and numerical information. These inputs provide the raw material that AI systems analyze during training. By studying large amounts of data, AI models begin to identify patterns, relationships, and structures that allow them to make predictions and generate responses.

The quality and diversity of data play a critical role in how well an AI system performs. Accurate, balanced, and relevant data helps models produce reliable results, while poor or biased data can lead to incorrect or unfair outcomes. This is why data collection, preparation, and evaluation are essential steps in the development of responsible AI systems.

The Data Family works closely with the Parameter Family, which adjusts internal values based on data, and the Neural Network Family, which processes and organizes data through layers of computation. It also supports the Token Family when dealing with language, as text data is converted into tokens for analysis.

Understanding the Data Family helps students recognize that AI does not "know" things on its own. Instead, it learns from the information it is given. The more meaningful and well-prepared the data, the more capable the AI system becomes.

In simple terms, data is the source of knowledge, and it is the starting point for all artificial intelligence.

## Data Family Breakdown Chart

| Component | Role | Description | Example |
|---|---|---|---|
| Data | Learning Material | Raw information used by AI systems to learn patterns and relationships | Text, images, audio, numbers |
| Dataset | Organized Collection | A structured group of data used for training and evaluation | A folder of labeled images |
| Training Data | Learning Input | Data used to teach the AI model during training | Thousands of sentences for a language model |
| Validation Data | Performance Check | Data used to tune the model during training without directly learning from it | Helps prevent overfitting |
| Test Data | Final Evaluation | Data used to measure how well the model performs after training | New unseen examples |
| Structured Data | Organized Format | Data arranged in tables with clear categories | Spreadsheets, databases |
| Unstructured Data | Flexible Format | Data without a predefined structure | Text documents, images, videos |
| Labeled Data | Guided Learning | Data tagged with correct answers to guide training | Image labeled "cat" or "dog" |
| Unlabeled Data | Self-Learning Input | Data without labels, used in unsupervised learning | Raw text without categories |
| Data Quality | Accuracy Measure | The reliability and correctness of data | Clean, complete, and consistent data |
| Data Bias | Imbalance Issue | When data does not represent all groups fairly | Skewed or incomplete datasets |
| Data Preprocessing | Preparation Step | Cleaning and organizing data before training | Removing errors, formatting text |

The Data Family reminds us that artificial intelligence is only as strong as the information it learns from. High-quality data leads to meaningful insights, while poor data can limit or misguide intelligent systems.

## Memory Family Introduction

The Memory Family represents how artificial intelligence systems store, retain, and retrieve information. Just as humans rely on memory to recall past experiences, learn new concepts, and make decisions, AI systems depend on memory to access previously processed information and maintain context.

In artificial intelligence, memory is not a single component but a collection of mechanisms that allow systems to handle information over time. Some forms of memory are short-term, holding information temporarily while a task is being completed. Other forms are long-term, storing learned knowledge that can be used across different tasks and interactions.

Memory plays a critical role in enabling AI systems to understand sequences, maintain context in conversations, and improve decision-making. For example, when a user interacts with a language model, the system uses memory to keep track of previous words, sentences, or prompts in order to generate coherent and relevant responses.

The Memory Family works closely with the Data Family, which provides the information to be stored, and the Parameter Family, which encodes learned knowledge within the model. It also supports the Neural Network Family by enabling systems to process sequences and retain important patterns over time.

Understanding the Memory Family helps students see that AI is not simply reacting to individual inputs in isolation. Instead, it relies on stored information and contextual awareness to produce meaningful and consistent outputs.

In simple terms, if data is what AI learns from, memory is how AI remembers and uses what it has learned.

## Memory Family Breakdown Chart

| Component | Role | Description | Example |
|---|---|---|---|
| Memory | Information Storage | The ability of an AI system to store and access information | Retaining previous inputs in a task |
| Short-Term Memory | Temporary Storage | Holds information for immediate processing | Keeping track of words in a sentence |
| Long-Term Memory | Persistent Storage | Stores learned knowledge for future use | Knowledge embedded in model parameters |
| Context Memory | Conversation Tracking | Maintains information within a specific interaction | Remembering earlier parts of a prompt |
| Working Memory | Active Processing | Handles information currently being used in computation | Processing a sequence step by step |
| External Memory | Extended Storage | Memory stored outside the model for retrieval | Databases, vector stores |
| Internal Memory | Embedded Knowledge | Information stored within the model itself | Learned patterns inside neural networks |
| Sequence Memory | Order Awareness | Tracks the order of inputs over time | Understanding sentence structure |
| Retrieval Mechanism | Access System | Retrieves stored information when needed | Searching relevant past data |

| Component | Role | Description | Example |
| --- | --- | --- | --- |
| Attention Mechanism | Focus System | Highlights important parts of memory for processing | Focusing on key words in a sentence |
| Memory Capacity | Storage Limit | The amount of information the system can handle at once | Context window size |

The Memory Family shows that intelligence is not just about processing information in the moment, but about retaining, organizing, and using knowledge over time to create meaningful understanding.

## Knowledge Family Introduction

The Knowledge Family represents how artificial intelligence organizes, connects, and uses information to create understanding. While data provides raw information and memory stores it, knowledge is what gives that information meaning and structure.

In artificial intelligence, knowledge is not simply a collection of facts. It is the result of identifying relationships, patterns, and connections within data. AI systems use knowledge to recognize concepts, make decisions, and generate meaningful responses. This allows machines to move beyond simple data processing and toward more intelligent behavior.

Knowledge can be represented in different ways within AI systems. Some forms are structured, such as databases and knowledge graphs, where relationships between concepts are clearly defined. Other forms are learned implicitly through neural networks, where patterns are encoded within parameters and used to guide predictions.

The Knowledge Family works closely with the Data Family, which provides the information needed to build knowledge, and the Memory Family, which stores and retrieves that information. It also connects with the Parameter Family, where learned knowledge is embedded within the model, and the Neural Network Family, which processes and organizes that knowledge.

Understanding the Knowledge Family helps students recognize that AI systems do not simply memorize information. Instead, they learn how pieces of information relate to one another, allowing them to interpret meaning, solve problems, and respond intelligently.

In simple terms, if data is information and memory stores it, knowledge is the understanding that makes it useful.

### Knowledge Family Breakdown Chart

| Component | Role | Description | Example |
|---|---|---|---|
| Knowledge | Understanding System | Organized information that allows AI to interpret meaning and make decisions | Recognizing relationships between concepts |
| Facts | Basic Information | Individual pieces of information stored within a system | "Water freezes at 0°C" |
| Relationships | Connection Mapping | Links between pieces of information that create meaning | "Teacher teaches student" |
| Knowledge Representation | Structure System | Methods used to organize and store knowledge in AI | Knowledge graphs, semantic networks |
| Knowledge Graph | Network Mapping | A structured representation of entities and their relationships | Google Knowledge Graph |
| Semantic Understanding | Meaning Interpretation | The ability to understand meaning behind words and data | Understanding synonyms and context |
| Inference | Reasoning Process | Drawing conclusions based on known information | If A = B and B = C, then A = C |
| Rules | Decision Framework | Logical guidelines used to make decisions | If temperature < 0°C → freeze |
| Ontology | Concept Framework | A structured system defining categories and relationships | Classification of animals |
| Explicit Knowledge | Direct Information | Clearly defined and stored knowledge | Rules, databases |
| Implicit Knowledge | Learned Patterns | Knowledge learned through experience and data | Patterns learned by neural networks |
| Knowledge Integration | System Coordination | Combining information from multiple sources | Merging text, images, and data |

The Knowledge Family transforms information into understanding, allowing artificial intelligence to move beyond data and memory into meaningful reasoning and intelligent decision-making.

## AI Model Family Introduction

The AI Model Family represents the core system within artificial intelligence that performs tasks such as prediction, classification, and content generation. It is the part of AI that users interact with, whether they are asking questions, generating images, or analyzing data.

An AI model is a trained system that has learned patterns from data. During training, the model processes large amounts of information and adjusts its internal parameters to improve accuracy.

Once trained, the model can take new input and produce meaningful output based on what it has learned.

AI models can take many forms depending on their purpose. Some models are designed to recognize images, others to understand language, and others to make decisions or predictions. Large Language Models (LLMs), for example, are designed to process and generate human language by predicting sequences of tokens.

The AI Model Family works closely with the Data Family, which provides the learning material, and the Parameter Family, which stores learned patterns. It also relies on the Neural Network Family for structure and the Chip and Infrastructure Families for execution and deployment.

Understanding the AI Model Family helps students see that AI is not just a concept, but a working system that has been trained to perform specific tasks. It is the engine that transforms input into output.

In simple terms, if AI is the system, the model is the part that actually does the thinking and producing.

## AI Model Family Breakdown Chart

| Component | Role | Description | Example |
|---|---|---|---|
| AI Model | Core System | The trained system that performs tasks | Chatbot, image generator |
| Training | Learning Process | Process of teaching the model using data | Learning from datasets |
| Inference | Output Generation | Using the trained model to produce results | Answering a question |
| Model Architecture | Structural Design | Defines how the model is built | Neural networks, transformers |
| Parameters | Learned Values | Internal values adjusted during training | Billions of weights |
| Input Data | Entry Information | Data provided to the model for processing | User prompt |
| Output Data | Result | The response or prediction generated | AI-generated text |
| Model Types | Functional Categories | Different kinds of models for different tasks | Classification, generation |
| Fine-Tuning | Model Adjustment | Improving a model for specific tasks | Custom-trained AI |
| Pretrained Model | Base System | A model trained on large datasets before use | General-purpose AI |

| Component | Role | Description | Example |
|---|---|---|---|
| Evaluation | Performance Check | Measuring how well the model performs | Accuracy, loss metrics |

The AI Model Family represents the working core of artificial intelligence, where learned patterns are transformed into meaningful actions, predictions, and responses.

## Neural Network Family Introduction

The Neural Network Family represents the structural foundation of how artificial intelligence learns and processes information. Inspired by the human brain, neural networks are systems of interconnected units, often called neurons, that work together to analyze data, recognize patterns, and make decisions.

In artificial intelligence, a neural network is made up of layers. The input layer receives data, the hidden layers process that data, and the output layer produces a result. As information flows through these layers, the network applies mathematical transformations that allow it to detect patterns and relationships within the data.

Neural networks rely heavily on the Parameter Family, where weights and biases determine how strongly information flows between neurons. During training, these parameters are adjusted to improve the network's performance. The more the network learns, the better it becomes at recognizing patterns and making accurate predictions.

The Neural Network Family works closely with the Data Family, which provides the information needed for learning, and the Memory and Knowledge Families, which help store and organize learned information. It also supports the Token Family when processing language, allowing AI systems to interpret sequences of tokens in meaningful ways.

Understanding neural networks helps students see that AI is not simply following instructions. Instead, it is learning from examples and improving over time through layered processing and pattern recognition.

In simple terms, neural networks are the systems that allow AI to think, learn, and make sense of information.

## Neural Network Family Breakdown Chart

| Component | Role | Description | Example |
|---|---|---|---|
| Neural Network | Learning Structure | A system of interconnected neurons that processes data and learns patterns | Image recognition system |

| Component | Role | Description | Example |
|---|---|---|---|
| Neuron (Node) | Processing Unit | A basic unit that receives input, processes it, and passes output forward | A single calculation point |
| Input Layer | Data Entry Point | The first layer that receives raw data | Pixels of an image |
| Hidden Layers | Processing Layers | Intermediate layers that transform data and detect patterns | Feature detection layers |
| Output Layer | Result Generator | The final layer that produces predictions or decisions | "Cat" vs "Dog" classification |
| Weights | Signal Strength | Values that determine how strongly inputs influence outputs | Higher weight = stronger impact |
| Bias | Adjustment Value | Helps shift outputs to improve learning accuracy | Fine-tuning predictions |
| Activation Function | Decision Function | Determines whether a neuron should activate based on input | ReLU, Sigmoid |
| Forward Propagation | Data Flow | The process of passing input data through the network to produce output | Input → Hidden → Output |
| Backpropagation | Learning Process | Adjusts weights and biases based on error to improve performance | Correcting mistakes |
| Loss Function | Error Measurement | Calculates how far the output is from the correct answer | Prediction vs actual |
| Deep Neural Network | Advanced Structure | A network with many hidden layers for complex learning | Deep learning models |

The Neural Network Family provides the structure that allows artificial intelligence to transform data into understanding, making it possible for machines to learn from experience and improve over time.

## AI Translator Architecture Family Introduction

The AI Translator Architecture Family represents the system that transforms input into meaningful output in artificial intelligence. It acts as the bridge between what humans provide and what AI produces. Whether a user types a question, uploads an image, or speaks a command, this architecture is responsible for interpreting that input and generating a response.

In artificial intelligence, translation does not only mean converting one language to another. It refers to the broader process of transforming one form of information into another. For example, text can be translated into a response, speech can be converted into text, and images can be interpreted into descriptions. This transformation process is at the core of how modern AI systems function.

The AI Translator Architecture is built on multiple foundational families. It relies on the Token Family to break input into manageable units, the Neural Network Family to process patterns, and the Parameter Family to guide learning. It also uses knowledge and memory to maintain context and produce coherent results.

At the center of this architecture is the ability to encode and decode information. Encoding transforms input into a format the model can understand, while decoding converts processed information back into human-readable output. This continuous transformation allows AI systems to interact with users in a natural and meaningful way.

Understanding the AI Translator Architecture Family helps students see that AI is not simply responding randomly. Instead, it is systematically translating input into output through structured processes and learned patterns.

In simple terms, if tokens are the language pieces and neural networks are the brain, the AI Translator Architecture is the system that turns understanding into communication.

## AI Translator Architecture Family Breakdown Chart

| Component | Role | Description | Example |
|---|---|---|---|
| AI Translator Architecture | Transformation System | Converts input into meaningful output | Question → Answer |
| Input Processing | Data Intake | Receives and prepares user input for analysis | Text prompt entered by a user |
| Encoding | Input Conversion | Transforms input into numerical representations | Words → token IDs |
| Context Building | Meaning Formation | Organizes input into a structured understanding | Sentence context tracking |
| Sequence Processing | Order Handling | Maintains the correct order of information | Word sequence in a sentence |
| Attention Mechanism | Focus System | Identifies important parts of input data | Highlighting key words |
| Representation Layer | Feature Mapping | Converts data into internal representations for processing | Embeddings |
| Decoding | Output Generation | Converts processed data back into human-readable form | Token IDs → words |
| Output Generation | Response Creation | Produces the final answer or result | AI-generated response |

| Component | Role | Description | Example |
|---|---|---|---|
| Multimodal Translation | Cross-Format Processing | Translates between different types of data | Image → text description |
| Feedback Loop | Improvement Cycle | Uses results to refine future outputs | Model learning over time |

The AI Translator Architecture Family shows that artificial intelligence is not just about understanding information, but about transforming it into meaningful communication that humans can use and interact with.

## AI Architecture Family Introduction

The AI Architecture Family represents the overall design and structure of an artificial intelligence system. While individual families such as Data, Parameters, and Neural Networks explain how AI learns and processes information, the AI Architecture Family shows how all these components are organized and work together as a complete system.

In artificial intelligence, architecture refers to the blueprint that defines how data flows through a system, how components interact, and how decisions are produced. It determines how input is received, how it is processed, and how output is generated. A well-designed architecture allows AI systems to operate efficiently, scale to large problems, and produce reliable results.

AI architectures can vary depending on the type of system being built. Some are designed for image recognition, others for language processing, and others for decision-making or automation. Despite these differences, most architectures share common elements such as input layers, processing layers, memory components, and output mechanisms.

The AI Architecture Family connects all foundational families. It integrates the Data Family for input, the Token Family for language processing, the Neural Network Family for computation, the Parameter Family for learning, and the Memory and Knowledge Families for storing and organizing information. It also works closely with the AI Translator Architecture Family to transform input into meaningful output.

Understanding the AI Architecture Family helps students see the “big picture” of artificial intelligence. Instead of viewing AI as separate parts, they begin to understand it as a coordinated system where each component plays a specific role.

In simple terms, if individual families are the parts of AI, the AI Architecture Family is the blueprint that brings everything together into one working system.

### AI Architecture Family Breakdown Chart

| Component | Role | Description | Example |
|---|---|---|---|
| AI Architecture | System Blueprint | The overall design that defines how an AI system is structured and operates | ChatGPT system design |
| Input Layer | Data Entry Point | Receives raw input from users or external sources | Text prompt, image upload |
| Preprocessing Layer | Data Preparation | Cleans and organizes input data before processing | Tokenization, normalization |
| Processing Core | Computation Engine | The main system where data is analyzed and transformed | Neural networks, transformers |
| Parameter System | Learning Core | Stores adjustable values that guide learning and predictions | Model weights and biases |
| Memory System | Context Storage | Maintains and retrieves information during processing | Context window, external memory |
| Knowledge System | Understanding Layer | Organizes relationships and meaning within the system | Knowledge graphs, learned patterns |
| Decision Layer | Output Logic | Determines the final output based on processed data | Selecting the best response |
| Output Layer | Result Delivery | Presents results to the user in a usable format | Text, image, or audio response |
| Feedback Loop | Improvement System | Uses outcomes to refine future performance | Model updates and retraining |
| Scalability System | Expansion Capability | Allows the system to handle increasing data and complexity | Cloud-based AI systems |
| Integration Layer | System Connection | Connects AI with external tools and platforms | APIs, databases, applications |

The AI Architecture Family reveals that artificial intelligence is not a single technology, but a coordinated system of interconnected components working together to transform data into intelligent outcomes.

## Chip Family Introduction

The Chip Family represents the physical hardware that powers artificial intelligence systems. While software components such as data, algorithms, and neural networks define how AI works, none of these systems can function without the computing chips that perform the actual calculations.

Chips are specialized electronic components designed to process information at extremely high speeds. In artificial intelligence, these chips handle the complex mathematical operations required for training models and generating outputs. Every prediction, calculation, and response produced by an AI system is executed by hardware within the Chip Family.

Different types of chips are used depending on the needs of the system. Central Processing Units (CPUs) handle general-purpose tasks, while Graphics Processing Units (GPUs) are optimized for parallel processing, making them ideal for training neural networks. More specialized chips, such as Tensor Processing Units (TPUs) and Neural Processing Units (NPUs), are designed specifically for AI workloads, enabling faster and more efficient computation.

The Chip Family works closely with the Neural Network Family, which defines the structure of computation, and the Parameter Family, which stores the values being processed. It also connects to the Infrastructure Family, where large-scale systems of chips are organized in data centers and cloud environments.

Understanding the Chip Family helps students recognize that artificial intelligence is not purely abstract. It depends on physical machines that require power, resources, and engineering to operate. These chips are the engines that bring AI models to life.

In simple terms, if AI is the intelligence, chips are the machines that make that intelligence possible.

## Chip Family Breakdown Chart

| Component | Role | Description | Example |
|---|---|---|---|
| Chip (Processor) | Computation Engine | Electronic hardware that performs calculations required for AI | CPU, GPU |
| CPU (Central Processing Unit) | General Processor | Handles a wide range of computing tasks | Running basic applications |
| GPU (Graphics Processing Unit) | Parallel Processor | Processes many calculations at once, ideal for AI training | Training neural networks |
| TPU (Tensor Processing Unit) | AI Accelerator | Specialized chip designed for machine learning tasks | Google TPU |
| NPU (Neural Processing Unit) | Edge AI Processor | Optimized for running AI on devices like phones | Smartphone AI features |
| AI Accelerator | Performance Booster | Hardware designed to speed up AI computations | Dedicated AI chips |
| Memory (Hardware) | Data Storage | Stores data and instructions for processing | RAM, VRAM |
| Parallel Processing | Speed Mechanism | Ability to perform multiple calculations simultaneously | GPUs processing thousands of operations |

| Component | Role | Description | Example |
|---|---|---|---|
| Throughput | Processing Capacity | Amount of data processed over time | High-performance computing systems |
| Latency | Response Time | Time it takes to process a request | Faster chips = lower latency |
| Power Consumption | Energy Usage | Amount of energy required to run the chip | High-performance GPUs use more power |
| Edge Devices | Local Processing | Devices that run AI locally without cloud support | Phones, smart cameras |

The Chip Family reminds us that behind every intelligent system is powerful hardware performing millions or billions of calculations, turning abstract models into real-world functionality.

## Watts Family Introduction

The Watts Family represents the measurement of power used by artificial intelligence systems. While the Chip Family provides the hardware that performs computations, the Watts Family explains how much energy those systems consume while operating.

A watt is a unit of power that measures the rate at which energy is used. In artificial intelligence, watts help us understand how much electricity is required to run processors, train models, and generate responses. Every time an AI system processes data, performs calculations, or produces output, it consumes power.

Different AI systems require different levels of power depending on their size and complexity. Small systems, such as those running on mobile devices, use relatively low amounts of power. In contrast, large-scale AI models operating in data centers can require significant amounts of electricity, especially during training, where billions of calculations are performed continuously.

The Watts Family works closely with the Chip Family, which determines how efficiently computations are performed, and the Infrastructure Family, where large networks of machines operate together. It also connects to the overall design of AI systems, as more efficient models can reduce power consumption while maintaining performance.

Understanding the Watts Family helps students recognize that artificial intelligence is not only about intelligence and computation, but also about energy use and resource management. As AI continues to grow, managing power efficiently becomes an important part of building responsible and sustainable systems.

In simple terms, if chips perform the work, watts measure how much power it takes to do that work.

### Watts Family Breakdown Chart

| Component | Role | Description | Example |
|---|---|---|---|
| Watt (W) | Power Measurement | Unit that measures the rate of energy use | A device using 100 watts |
| Power Consumption | Energy Usage Rate | Amount of power used during operation | AI model running on GPUs |
| Energy Efficiency | Performance Balance | How effectively a system uses power to perform tasks | More output with less power |
| High-Performance Computing | Intensive Power Use | Systems that require large amounts of power for complex tasks | AI training clusters |
| Idle Power | Standby Usage | Power consumed when systems are not actively processing | Servers waiting for requests |
| Peak Power | Maximum Usage | Highest level of power used during heavy workloads | Training large AI models |
| Thermal Output | Heat Generation | Heat produced as a result of power usage | Cooling systems in data centers |
| Cooling Systems | Temperature Control | Systems used to manage heat from high power usage | Air or liquid cooling |
| Power Supply | Energy Source | Provides electricity to AI systems | Electrical grid, batteries |
| Edge Power Usage | Local Energy Use | Power consumption on smaller devices | Smartphones, IoT devices |
| Data Center Power | Large-Scale Usage | Power required for large AI operations | Server farms running AI models |
| Sustainable Energy | Efficiency Goal | Use of renewable energy to reduce environmental impact | Solar-powered data centers |

The Watts Family highlights that intelligence comes with a cost—every AI system requires power, and understanding that power is essential for building efficient and sustainable technologies.

## ⚡ Watts Family (AI Power Usage Scale)

### Definition

The **Watts Family** represents the real-time power usage of artificial intelligence systems, measured in watts (W). It shows how much power AI systems consume at a given moment as they process data and perform computations.

## Watts Scale

1 Watt
10 Watts
100 Watts
500 Watts
1,000 Watts (1 Kilowatt)
10,000 Watts (10 Kilowatts)
100,000 Watts (100 Kilowatts)
1,000,000 Watts (1 Megawatt)
10,000,000 Watts (10 Megawatts)
100,000,000+ Watts (100+ Megawatts)

## Key Characteristics

Measured in real-world electrical power

Higher watt values indicate higher processing demand

Used to represent the scale of AI systems from small devices to large infrastructures

Closely connected to the Chip Family, Energy Family, and Infrastructure Family

## Examples

Low watts → small devices and basic AI functions

Medium watts → computers and advanced applications

High watts → data centers and large AI systems

## Summary

The **Watts Family** provides a numerical representation of how much power artificial intelligence systems use in real time, helping to illustrate the scale and intensity of AI operations.

## 🧠 Student Insight

*Watts show how much power AI is using right now.*

## AI Energy Family Introduction

The AI Energy Family represents the total amount of energy consumed by artificial intelligence systems over time. While the Watts Family measures the rate at which power is used at any given moment, the Energy Family focuses on the accumulated energy required to perform tasks, run systems, and sustain operations.

In artificial intelligence, energy is used every time a model is trained, a system processes data, or a response is generated. Large-scale AI systems, especially those operating in data centers, can consume significant amounts of energy due to the continuous processing of massive datasets and complex computations.

Energy is typically measured over time using units such as kilowatt-hours (kWh), which reflect how much power has been used during a specific period. For example, running a high-performance AI system for several hours or days results in a total energy cost that goes beyond the moment-to-moment power usage measured in watts.

The AI Energy Family works closely with the Watts Family, which provides the rate of power consumption, and the Chip Family, which determines how efficiently computations are performed. It also connects to the Infrastructure Family, where large-scale systems operate continuously and require long-term energy management.

Understanding the AI Energy Family helps students recognize that artificial intelligence has real-world resource implications. It is not only about performance and speed, but also about sustainability, efficiency, and responsible use of technology.

In simple terms, if watts measure how fast energy is used, energy measures how much is used over time.

## AI Energy Family Breakdown Chart

| Component | Role | Description | Example |
|---|---|---|---|
| Energy | Total Consumption | The total amount of power used over a period of time | Running an AI system for hours or days |
| Kilowatt-hour (kWh) | Energy Measurement | Unit used to measure energy consumption over time | 1 kWh = using 1,000 watts for 1 hour |
| Energy Usage | Consumption Tracking | Total energy required for AI operations | Training a large model |
| Training Energy | Learning Cost | Energy used during model training processes | Weeks of GPU usage |
| Inference Energy | Response Cost | Energy used when AI generates outputs | Answering a user prompt |

| Component | Role | Description | Example |
|---|---|---|---|
| Energy Efficiency | Optimization Goal | Reducing energy use while maintaining performance | Efficient AI models |
| Energy Scaling | Growth Impact | Increase in energy use as systems grow larger | Bigger models = more energy |
| Data Center Energy | Infrastructure Demand | Total energy used by large AI facilities | Server farms operating 24/7 |
| Cooling Energy | Temperature Control | Energy required to cool systems and prevent overheating | Air conditioning in data centers |
| Renewable Energy | Sustainability Source | Use of clean energy to power AI systems | Solar, wind-powered data centers |
| Carbon Impact | Environmental Effect | Emissions associated with energy consumption | AI training carbon footprint |
| Energy Management | Resource Control | Strategies to monitor and reduce energy use | Efficient scheduling, hardware optimization |

The AI Energy Family reminds us that artificial intelligence operates within the physical world, where every computation consumes energy and every system has an impact on resources and sustainability.

## Energy Family (Total Power Consumption of Artificial Intelligence)

### Definition

The **Energy Family** represents the total amount of electricity used by artificial intelligence systems over time. It is measured in watt-hours (Wh) or kilowatt-hours (kWh) and shows how much energy AI systems consume while running.

### Energy Scale

1 Watt-hour (Wh)
10 Watt-hours (Wh)
100 Watt-hours (Wh)
500 Watt-hours (Wh)
1,000 Watt-hours (1 Kilowatt-hour, kWh)
10,000 Watt-hours (10 kWh)
100,000 Watt-hours (100 kWh)
1,000,000 Watt-hours (1 Megawatt-hour, MWh)
10,000,000 Watt-hours (10 MWh)
100,000,000+ Watt-hours (100+ MWh)

### Key Characteristics

Measured over time (not instant like watts)

Represents total electricity consumption

Increases the longer AI systems run

Directly related to cost and energy usage

Closely connected to Watts Family, Chip Family, and Infrastructure Family

### Examples

Low energy → short AI tasks or small devices

Medium energy → daily AI usage on computers

High energy → training AI models and running data centers

### Summary

The **Energy Family** provides a numerical representation of the total electricity consumed by artificial intelligence systems over time, helping to illustrate the overall cost and impact of AI operations.

### 🧠 Student Insight

*Energy shows how much total power AI has used over time.*

## AI Infrastructure Family Introduction

The AI Infrastructure Family represents the large-scale systems and environments that support, power, and connect artificial intelligence technologies. While individual components such as chips, data, and models explain how AI works, infrastructure explains where and how these systems operate in the real world.

AI infrastructure includes data centers, cloud platforms, networking systems, and storage environments that allow artificial intelligence to function at scale. These systems provide the computing power, data access, and connectivity needed to train models, process information, and deliver results to users around the world.

Modern AI systems rely heavily on cloud-based infrastructure, where thousands of machines work together to handle massive workloads. This allows AI to scale efficiently, making it possible to process large datasets, run complex models, and serve millions of users simultaneously. Infrastructure also ensures reliability, security, and continuous availability of AI services.

The AI Infrastructure Family works closely with the Chip Family, which provides the hardware, the Watts and Energy Families, which manage power consumption, and the Data Family, which supplies the information being processed. It also supports the AI Architecture Family by providing the environment where all system components are deployed and connected.

Understanding the AI Infrastructure Family helps students recognize that artificial intelligence is not contained within a single device. Instead, it operates across global systems that require coordination, resources, and engineering to function effectively.

In simple terms, if AI is the system and chips are the engines, infrastructure is the environment that allows everything to run at scale.

## AI Infrastructure Family Breakdown Chart

| Component | Role | Description | Example |
|---|---|---|---|
| AI Infrastructure | System Environment | The physical and digital systems that support AI operations | Global AI platforms |
| Data Center | Processing Hub | Facilities that house servers and computing equipment | Large server farms |
| Cloud Computing | Scalable Platform | Remote systems that provide computing resources over the internet | AWS, Azure, Google Cloud |
| Servers | Compute Units | Machines that process data and run AI models | Rack-mounted servers |
| Networking | Connectivity System | Systems that allow communication between machines | Internet, fiber networks |
| Storage Systems | Data Management | Systems that store large volumes of data | Databases, cloud storage |
| Distributed Computing | Workload Sharing | Splitting tasks across multiple machines for efficiency | Parallel processing across clusters |
| Edge Infrastructure | Local Processing | Running AI closer to the user or device | Smart devices, IoT systems |
| Load Balancing | Traffic Control | Distributes workloads to prevent overload | Managing user requests |
| Security Systems | Protection Layer | Safeguards data and systems from threats | Encryption, firewalls |
| Redundancy | Reliability System | Backup systems to ensure continuous operation | Failover servers |

| Component | Role | Description | Example |
|---|---|---|---|
| Scalability | Growth Capability | Ability to expand resources as demand increases | Adding more servers dynamically |

The AI Infrastructure Family reveals that artificial intelligence operates on a global scale, relying on interconnected systems that provide the power, storage, and connectivity needed to support intelligent technologies.

## AI Agent Family Introduction

The AI Agent Family represents systems that can take action, make decisions, and perform tasks on behalf of users or other systems. Unlike traditional AI models that only respond to input, AI agents are designed to operate with a level of autonomy, allowing them to plan, execute, and adapt to achieve specific goals.

An AI agent receives input from its environment, processes that information, and takes actions based on predefined objectives or learned behaviors. These actions can include answering questions, automating workflows, retrieving information, or interacting with other systems. Some agents operate in simple environments, while others function in complex systems that require continuous decision-making.

AI agents often combine multiple foundational families. They use the Data Family to gather information, the Memory and Knowledge Families to retain and understand context, and the Neural Network and Parameter Families to process and learn from interactions. They also rely on the AI Architecture and Infrastructure Families to operate reliably at scale.

Modern AI agents can work independently or as part of larger systems. For example, a customer support agent can respond to inquiries automatically, while a more advanced agent can complete multi-step tasks such as scheduling, research, or system management.

Understanding the AI Agent Family helps students see that artificial intelligence is not only about generating responses, but also about taking meaningful action in real-world scenarios.

In simple terms, if AI models think and respond, AI agents act and execute.

## AI Agent Family Breakdown Chart

| Component | Role | Description | Example |
|---|---|---|---|
| AI Agent | Action System | An AI system that can make decisions and perform tasks | Virtual assistant |

| Component | Role | Description | Example |
|---|---|---|---|
| Environment | Operating Context | The space where the agent interacts and gathers information | Web, apps, databases |
| Input Perception | Data Intake | Receives information from the environment | User query or sensor input |
| Decision Engine | Action Selection | Determines what action to take based on input | Choosing a response or task |
| Action Execution | Task Performance | Carries out decisions in the environment | Sending an email, retrieving data |
| Goal System | Objective Setting | Defines what the agent is trying to achieve | Complete a task or solve a problem |
| Planning Module | Strategy Builder | Breaks down tasks into steps | Multi-step problem solving |
| Feedback Loop | Learning Cycle | Improves performance based on outcomes | Adjusting actions over time |
| Autonomy Level | Independence Scale | Degree of independence in decision-making | Fully automated vs assisted |
| Multi-Agent System | Collaboration Network | Multiple agents working together | Coordinated AI systems |

## Ingredient Family Introduction

The Ingredient Family represents the essential components that come together to create an artificial intelligence system. Just as a recipe requires specific ingredients to produce a final dish, AI systems rely on a combination of foundational elements working together to function effectively.

Artificial intelligence is not built from a single component. It requires data to learn from, algorithms to guide processing, models to perform tasks, hardware to execute computations, and energy to power the entire system. Each of these elements plays a critical role, and the absence of any one component would prevent the system from operating properly.

The Ingredient Family provides a simplified way for students to understand AI as a system made up of interconnected parts. Instead of viewing AI as complex or mysterious, learners can begin to see it as something constructed from identifiable and understandable components.

This family connects directly with all other foundational families, including Data, Algorithm, Neural Network, Parameter, Chip, and Energy Families. It serves as an overview that brings these elements together into a unified perspective.

Understanding the Ingredient Family helps students recognize that artificial intelligence is built, not magical. It is the result of carefully combining the right components in the right way.

In simple terms, if AI is the final product, the Ingredient Family represents everything needed to build it.

### Ingredient Family Breakdown Chart

| Component | Role | Description | Example |
|---|---|---|---|
| Data | Learning Input | Provides the information AI uses to learn patterns | Text, images, audio |
| Algorithms | Instruction Set | Guides how data is processed and decisions are made | Sorting, classification |
| AI Models | Execution System | Performs tasks such as prediction or generation | Language models |
| Neural Networks | Processing Structure | Organizes how data is processed through layers | Deep learning models |
| Parameters | Learning Adjustments | Fine-tune how the model learns and makes predictions | Weights and biases |
| Tokens | Language Units | Breaks text into smaller pieces for processing | Words or subwords |
| Hardware (Chips) | Computation Engine | Executes calculations required by AI systems | GPUs, CPUs |
| Energy | Power Source | Supplies the energy needed to run systems | Electricity usage |
| Infrastructure | Support System | Provides the environment for AI to operate at scale | Cloud systems, data centers |
| Memory | Storage System | Stores and retrieves information for processing | Context memory |
| Knowledge | Understanding Layer | Represents meaning and relationships within data | Knowledge graphs |

The Ingredient Family shows that artificial intelligence is built from a combination of essential components, each playing a vital role in creating intelligent systems.

## The Frontend and Backend Family of Artificial Intelligence

### *How AI Systems Connect User Interaction to Intelligent Processing*

#### Introduction

Every artificial intelligence system operates through two essential layers: the frontend and the backend. These layers work together to transform human interaction into intelligent responses, allowing AI systems to function smoothly and effectively.

The frontend is the part of the system that users see and interact with. It includes screens, input tools, and the display of results. Whether a user is typing a question, speaking a command, or uploading an image, all interaction begins at the frontend.

Behind the scenes, the backend serves as the processing engine of the system. It receives input from the frontend, manages data, connects to the AI model, and performs the computations needed to generate a response. The backend is where the intelligence of the system operates, using models, algorithms, and infrastructure to interpret and respond to user requests.

This visual illustrates how these two layers work together as a complete system. It shows the flow of information from user input to AI processing and back to the user as a result. By understanding this structure, students can see that AI is not a single tool, but a coordinated system of layers working together.

In simple terms, the frontend is where interaction happens, and the backend is where intelligence is created.

***Every AI system has two sides: what you see and what actually does the thinking.***

**This structure helps students understand how every interaction with AI is supported by a hidden system of processing and intelligence.**

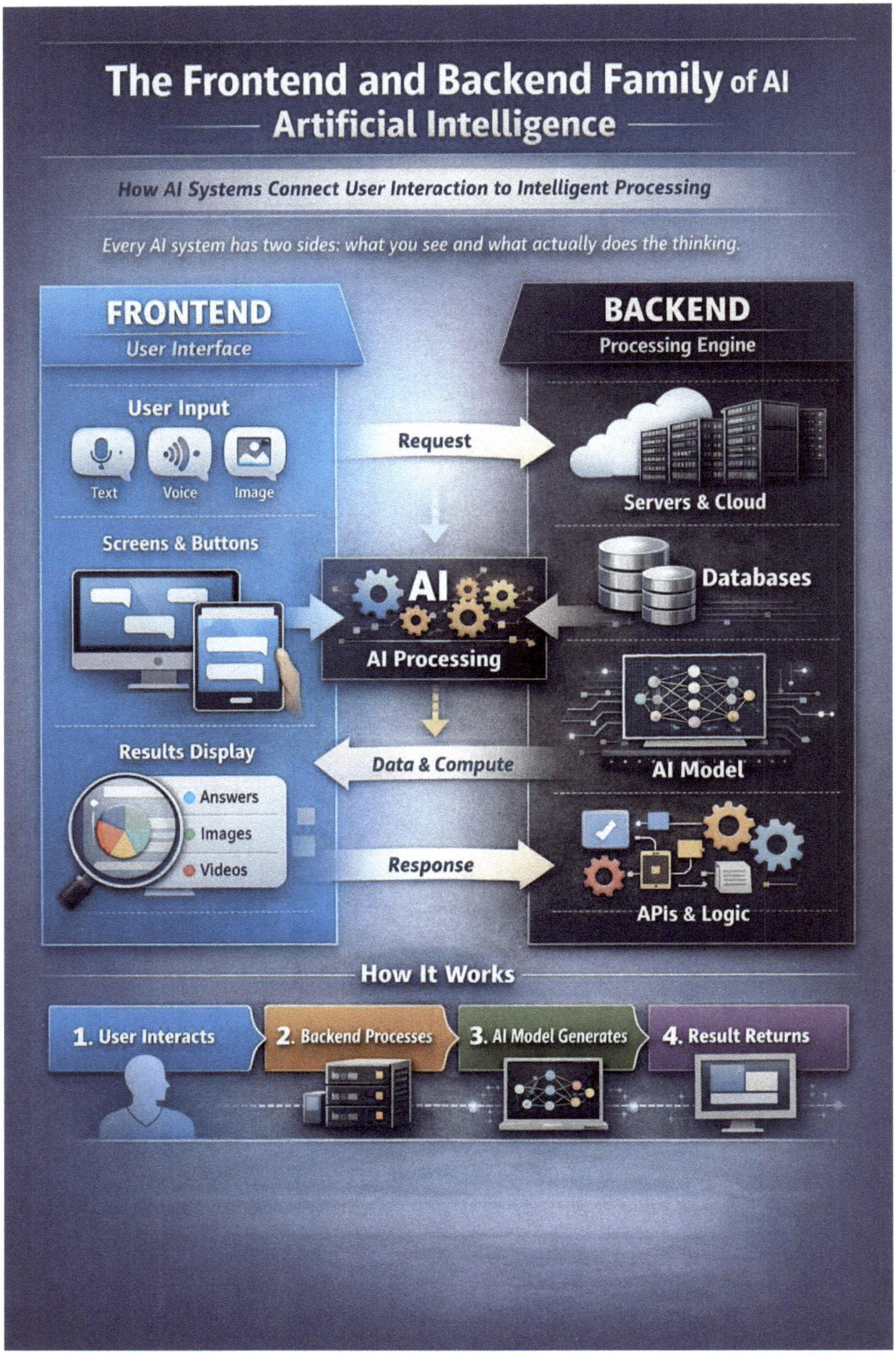
The Frontend and Backend Family of AI
Artificial Intelligence
How AI Systems Connect User Interaction to Intelligent Processing
Every AI system has two sides: what you see and what actually does the thinking.
FRONTEND
User Interface
User Input
Text
Voice
Image
Screens & Buttons
Results Display
Answers
Images
Videos
Request
AI
AI Processing
Data & Compute
Response
BACKEND
Processing Engine
Servers & Cloud
Databases
AI Model
APIs & Logic
How It Works
1. User Interacts
2. Backend Processes
3. AI Model Generates
4. Result Returns

## AI Qubit Family Introduction

The AI Qubit Family represents the foundation of quantum computing as it relates to artificial intelligence. While traditional computing relies on bits, which can exist as either 0 or 1, quantum computing uses qubits, which can exist in multiple states simultaneously.

A qubit is the basic unit of quantum information. Unlike classical bits, qubits can take advantage of properties such as superposition and entanglement, allowing quantum systems to process complex computations more efficiently in certain scenarios. This opens new possibilities for solving problems that are difficult or impossible for traditional computers.

In the context of artificial intelligence, qubits have the potential to accelerate learning processes, optimize large-scale systems, and improve complex simulations. Although quantum AI is still in development, it represents a future direction where computational power can expand beyond current limitations.

The AI Qubit Family connects with the Chip Family, as quantum processors are specialized hardware, and with the Algorithm Family, where new types of quantum algorithms are designed. It also relates to the Parameter and Neural Network Families, as researchers explore quantum-enhanced learning models.

Understanding the AI Qubit Family helps students see that artificial intelligence is continuously evolving. It introduces the idea that the foundations of computing themselves can change, leading to new forms of intelligence and problem-solving.

In simple terms, if bits are the foundation of today's computing, qubits represent the foundation of tomorrow's possibilities.

## AI Qubit Family Breakdown Chart

| Component | Role | Description | Example |
|---|---|---|---|
| Qubit | Quantum Unit | Basic unit of quantum information | Quantum bit in a quantum computer |
| Superposition | Multi-State Ability | Ability to exist in multiple states at once | 0 and 1 simultaneously |
| Entanglement | Connection Property | Linking qubits so they influence each other | Paired quantum states |
| Quantum Gate | Operation System | Performs operations on qubits | Quantum logic gates |
| Quantum Circuit | Processing Structure | Sequence of quantum operations | Quantum algorithms |
| Quantum Processor | Hardware Engine | Specialized chip for quantum computing | Quantum computer hardware |

| Component | Role | Description | Example |
|---|---|---|---|
| Quantum Algorithm | Computation Method | Algorithm designed for quantum systems | Optimization problems |
| Quantum Speedup | Performance Advantage | Faster processing for certain problems | Complex simulations |
| Quantum Noise | Stability Challenge | Errors caused by environmental interference | Qubit instability |
| Quantum Error Correction | Stability Solution | Techniques to reduce errors in quantum systems | Fault-tolerant computing |

The AI Qubit Family introduces a new frontier in computing, where the limits of traditional systems are expanded, opening the door to more powerful and advanced forms of artificial intelligence.

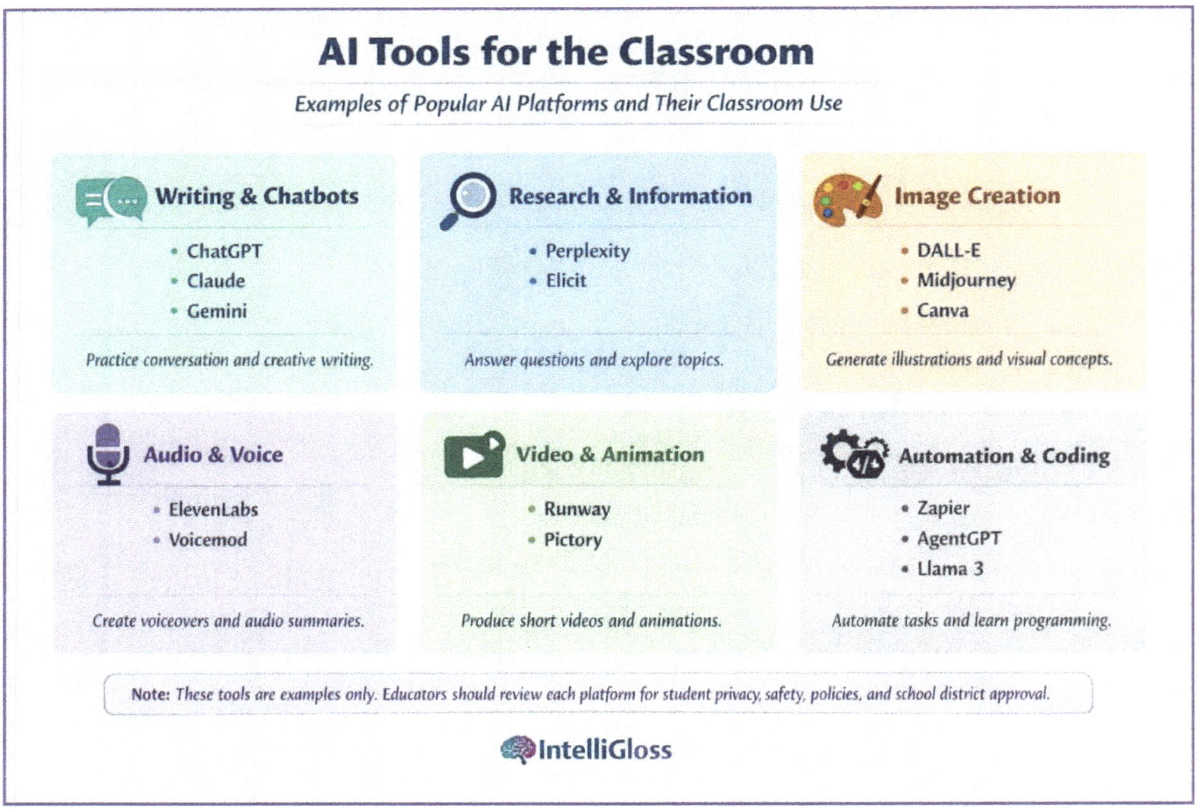

## How This Book Begins

Now that we understand why artificial intelligence matters, we begin with its foundation.

In Module 1, students will explore how information is represented, processed, and transformed into intelligence. This includes understanding the basic building blocks of computing and how data becomes meaningful output.

This foundation is essential, because every advanced AI system begins with simple principles.

## Introduction – From Atoms to Algorithms

### Opening Vision

Every civilization builds tools to understand its own mind.
For the 21st century, that tool is *Artificial Intelligence* — and at the root of every intelligent system lies the most powerful simplicity ever discovered: **binary code**.

Binary is more than just a language of *ones* and *zeros* — it is the **foundation of logic, learning, and digital consciousness**. Just as atoms form the matter that shapes our universe, binary digits form the logic that powers our machines, networks, and now, the very idea of synthetic intelligence.

When we explore the binary foundation of intelligence, we are not just learning how computers think — we are uncovering how *thinking itself* can be modeled, measured, and magnified.

### Learning Objectives

By the end of this introduction, students and readers will:

1. Understand that binary code represents the universal logic of all computing systems.
2. Recognize the historical path from natural patterns to digital logic.
3. See connections between binary operations and human reasoning.
4. Appreciate how binary thinking forms the bridge between science, mathematics, and artificial intelligence.
5. Develop curiosity for how binary shapes every digital experience — from smartphone screens to neural networks.

### The Birth of Binary Thinking

Before there were computers, there was thought — human thought.
Ancient philosophers sought to explain the world through **duality**: light and darkness, truth and falsehood, life and death. These opposites gave rise to understanding through contrast. Centuries

later, mathematician **Gottfried Wilhelm Leibniz** saw that numbers could express this same universal duality — through only **two symbols**, 1 and 0.

From that discovery, the digital age was born. Every calculation, every pixel, every stream of sound or image now passes through binary gates — like rivers of logic flowing through silicon valleys. Binary is the heartbeat of every algorithm, the rhythm that keeps our digital civilization alive.

## The Human Connection

Binary might seem mechanical, but it mirrors the way we think.
Humans constantly make **binary decisions** — yes or no, right or wrong, move or stay.
Our neurons fire in pulses that resemble digital on/off patterns.
In both the brain and the machine, *intelligence emerges through patterns of activation*.

Understanding binary, therefore, is not just computer science — it's **cognitive science**, **philosophy**, and **education** combined. It helps us ask:

- How do we define intelligence?
- Can reasoning be reduced to a series of logical choices?
- Where does emotion fit within a binary framework?

## From Logic to Learning

Every machine that learns, from a calculator to a self-driving car, begins its journey in binary.
A single "1" can mean **true**, **on**, or **signal present**, while "0" represents **false**, **off**, or **no signal**.
When combined through *logic gates* and *algorithms*, these simple symbols form vast systems of understanding — capable of solving equations, recognizing speech, and translating languages.

Just as a student starts with letters before mastering sentences, AI begins with bits before mastering knowledge.

## Why This Book Matters

This book, *The Binary Foundation of Intelligence*, is written for **students, teachers, and schools** who want to see beyond the screen — to understand **how thinking becomes digital**.
It aligns with modern **STEM and computer science standards** and introduces binary concepts in a way that connects math, physics, biology, and ethics.

Every chapter builds toward one goal: to reveal how binary logic shaped the birth of machines — and how it will continue shaping the minds of tomorrow.

## Reflection Questions

1. Why is binary called the foundation of all digital intelligence?
2. How does binary logic mirror human decision-making?

3. In what ways does understanding binary prepare us for the future of AI?
4. What can the simplicity of "1 and 0" teach us about the complexity of thought?

# Chapter 1 – The Language of Ones and Zeros

## Overview

Every idea in the digital age—every photo, song, word, or motion—can be expressed as a series of **1s and 0s**.
This chapter explores how **binary language** became the foundation of computing and how it mirrors the logical structure of human reasoning. Through binary, machines gain the ability to think, calculate, and learn — transforming pure electricity into *intelligence*.

## Learning Objectives

By the end of this chapter, students will be able to:

1. Define binary code and explain why it is essential to computing.
2. Describe how 1s and 0s represent logical states and data.
3. Compare binary decision-making in machines and human thought.
4. Understand how logic gates use binary to perform reasoning.
5. Recognize how binary systems underpin communication, storage, and artificial intelligence.

## 1.1 The Birth of a Digital Language

Binary code was born from simplicity — the idea that **complex systems can emerge from two basic states**: on or off, yes or no, 1 or 0.

Before computers, mathematicians like **Gottfried Wilhelm Leibniz (1646–1716)** envisioned a universal language of logic built upon two symbols. Leibniz saw the binary system as more than a mathematical curiosity — he believed it reflected the structure of reality itself.

Centuries later, engineers discovered that binary aligned perfectly with **electrical systems**, where a signal is either present (1) or absent (0). This harmony between **mathematics** and **electricity** gave rise to modern computing.

## 1.2 Understanding the Digits

At its core, binary code represents information using just two symbols.

- **1** means "on," "true," or "signal present."
- **0** means "off," "false," or "no signal."

Each binary digit is called a **bit**, short for *binary digit*.
While a single bit can represent only two possibilities, multiple bits combined can express millions of variations.

For example:

- 1 bit → 2 combinations (0, 1)
- 2 bits → 4 combinations (00, 01, 10, 11)
- 8 bits → 256 combinations (used in ASCII text and color values)

Every digital file—text, image, sound, or video—is encoded in this way.

## 1.3 The Power of Simplicity

Why does the entire digital universe rely on such simplicity?
Because **binary is reliable, efficient, and universal**.

Electric circuits cannot easily handle shades of "maybe." They need clear states: electricity is either flowing or it isn't. This clarity allows systems to **store, transmit, and process** information without confusion or loss.

In the human brain, neurons behave in a similar way — they either fire or rest. Though the brain's system is analog and biological, the pattern of activation resembles binary logic. Both systems prove that **clarity creates complexity**: from two choices, infinite structures arise.

## 1.4 From Bits to Bytes

A **byte** is a group of **8 bits**, capable of representing 256 possible combinations.
Each byte can encode a letter, a color tone, or a small unit of sound.

Examples:

- The letter **A** = 01000001
- The color **white** (RGB 255,255,255) = binary sequence of 24 bits
- A musical note stored in MP3 format = thousands of binary samples per second

Just as letters form words and words form sentences, **bits and bytes** form the digital language that drives artificial intelligence.

## 1.5 Logic in Action

Binary becomes truly intelligent when combined through **logic gates** — the building blocks of computation.

Common logic gates include:

- **AND Gate:** Output is 1 only if both inputs are 1.
- **OR Gate:** Output is 1 if either input is 1.
- **NOT Gate:** Flips the input — turning 1 into 0, or 0 into 1.

When arranged in circuits, these gates mimic the **neurons and synapses** of the human brain. Each decision, from simple addition to deep neural learning, is powered by billions of such logical operations occurring every second.

## 1.6 Binary and Human Thought

At first glance, binary logic seems mechanical — far removed from human creativity or emotion. Yet, deep within our cognition lies a similar rhythm.

Every decision we make relies on a binary comparison:
Should I go or stay?
Is this true or false?
Is the answer right or wrong?

Our brains filter information through **contrasts**, just like computers filter signals through binary states. The difference is that humans can interpret shades of meaning — a skill that AI continues to learn through **machine learning and probabilistic reasoning**.

## 1.7 Binary Communication

When two computers communicate, they don't speak English or Spanish — they exchange streams of binary signals known as **machine language**.
For example:

```
01001000 01100101 01101100 01101100 01101111
```

translates to **"Hello"** in ASCII code.

This silent dialogue of 1s and 0s happens billions of times per second across networks, powering emails, streaming, and AI models that learn from human input.

## 1.8 Binary in Artificial Intelligence

Artificial Intelligence begins with binary.
Every **neural network**, **algorithm**, and **data model** depends on binary logic to operate. When an AI system recognizes a face, classifies an image, or predicts an answer, it is processing patterns of binary values — comparing weights, thresholds, and outcomes.

In AI, binary forms the **skeleton of logic** upon which layers of complexity — such as learning, perception, and creativity — are built.

## 1.9 Why Binary Still Matters

Even in the age of quantum computing, binary remains fundamental.
Quantum bits (qubits) may hold multiple states, but when measured, they still collapse into a binary result — 0 or 1. This confirms that **binary logic is the universal foundation** for all digital reasoning, classical or quantum.

## Classroom Discussion Prompts

1. How do binary digits reflect the structure of human decision-making?
2. Why might simplicity (only two options) be better for machines than complexity?
3. How might understanding binary help us design fairer, smarter AI systems?
4. Can creativity exist within binary limitations? Discuss with examples.

# Chapter 1 – The Language of Ones and Zeros

## Chapter Summary

Binary code is the foundational language of all digital systems. Every form of digital information—text, images, sound, and video—is represented using combinations of two simple digits: 1 and 0. These digits represent logical states such as "on" and "off," forming the basis of how computers process and store information.

The binary system originated from early mathematical ideas, particularly those of Gottfried Wilhelm Leibniz, and later aligned perfectly with electrical systems. This alignment allowed machines to reliably process information using clear, distinct states.

As binary digits (bits) combine, they form larger units such as bytes, enabling complex data representation. Through logic gates such as AND, OR, and NOT, binary values are used to perform decision-making processes within computers.

Binary logic also parallels human thinking, where decisions are often based on comparisons such as true/false or yes/no. This connection highlights how simple systems can produce complex reasoning.

In artificial intelligence, binary serves as the underlying structure that supports algorithms, neural networks, and machine learning systems. Even with advancements like quantum computing, binary remains a fundamental component of digital reasoning.

## Reflection Section

### Student Reflection

Answer the following questions in complete sentences:

Why do you think binary uses only two digits instead of more complex numbers?

How does binary relate to the way humans make decisions?

What surprised you most about how computers use 1s and 0s?

In your own words, explain how simple systems can create complex results.

How might understanding binary help you better understand artificial intelligence?

## Worksheet: Understanding Binary Concepts

### Part A – Vocabulary Matching

Match the term with the correct definition:

A. Binary Code
B. Bit
C. Byte
D. Logic Gate
E. AND Gate

___ A single unit of binary data (0 or 1)

___ A system using 1s and 0s to represent data

___ A circuit that performs logical operations

___ A group of 8 bits

___ Outputs 1 only if both inputs are 1

### Part B – Short Answer

What does the number "1" represent in binary?

What does the number "0" represent in binary?

How many combinations can 2 bits create?

How many combinations can 8 bits create?

Why is binary considered reliable for computers?

**Part C – Thinking Exercise**

Give an example of a real-life decision that can be described in binary terms.

Explain how logic gates are similar to human decision-making.

## Quiz: Chapter 1 Assessment

**Multiple Choice**

Binary code uses which two digits?
A. 0 and 2
B. 1 and 10
C. 1 and 0
D. 2 and 4

A "bit" is:
A. A group of 8 numbers
B. A single binary digit
C. A type of computer
D. A logic gate

A byte consists of:
A. 2 bits
B. 4 bits
C. 8 bits
D. 16 bits

Which logic gate outputs 1 only if both inputs are 1?
A. OR
B. NOT
C. AND
D. XOR

Binary is important because it is:
A. Complicated

B. Random
C. Reliable and simple
D. Difficult to use

**Short Answer**

Explain one way binary is used in artificial intelligence.

What is one similarity between binary logic and human thinking?

## ✅ Answer Key

### Worksheet Answers

**Part A**

B

A

D

C

E

**Part B**
6. "On," "true," or signal present
7. "Off," "false," or no signal
8. 4 combinations
9. 256 combinations
10. Because it uses clear, distinct states (on/off), reducing errors

**Part C (Sample Answers)**
11. Example: Yes/No decision (go or stay, true or false)
12. Both involve making decisions based on conditions or inputs

### Quiz Answers

C

B

C

C

C

**Short Answer (Sample Responses)**
6. AI uses binary to process data, make predictions, and run algorithms
7. Both rely on decision-making processes such as true/false comparisons

# Chapter 2 – The Birth of Digital Thinking

## Overview

Before machines could think in binary, humans had to imagine that thought itself could be **measured**, **encoded**, and **replicated**.
The birth of digital thinking was not a single invention but a slow awakening — a realization that intelligence could exist outside the human mind.
This chapter explores how mathematical logic, engineering breakthroughs, and philosophical curiosity merged to create the first sparks of digital consciousness.

## Learning Objectives

By the end of this chapter, students will be able to:

1. Trace the historical timeline that led from early mathematics to modern computing.
2. Explain how mechanical reasoning evolved into programmable logic.
3. Identify key pioneers who shaped digital thought.
4. Understand how binary systems changed the human definition of intelligence.
5. Reflect on the social and ethical consequences of early computing revolutions.

## 2.1 The Seeds of Logic

Long before electricity lit up a single circuit, philosophers searched for the **rules of reasoning**.
Aristotle proposed *syllogisms*—if A implies B and B implies C, then A implies C.
This was the earliest form of **logical inference**, the same principle that drives every algorithm today.

Centuries later, **George Boole (1815–1864)** transformed these philosophical patterns into a mathematical system called **Boolean Logic**, where statements could be only *true* or *false*, 1 or 0.

Boole's equations became the grammar of digital thought — the invisible script that computers still follow.

## 2.2 From Symbols to Switches

When engineers began experimenting with electricity in the 19th century, they discovered that **electrical switches** could physically represent Boole's logical states.
A closed circuit (current flows) equaled 1.
An open circuit (no current) equaled 0.

This simple mapping united **abstract logic** with **physical reality**.
The telegraph, telephone relays, and later transistors all spoke the same silent binary language.
This moment marked the beginning of machines translating human logic into physical action.The world had entered the era where **ideas could flow as electrons**.

## 2.3 The Mechanical Dream

Before electronics, inventors like **Charles Babbage** designed mechanical engines to perform calculations.
His *Analytical Engine* (1837) used gears and levers instead of circuits, but it contained the blueprint of a modern computer — memory, processing, and conditional instructions.
Mathematician **Ada Lovelace** recognized its potential not just for numbers but for patterns, music, and art.
She imagined a machine that could "compose symphonies if properly instructed."

Lovelace's vision became the philosophical seed of artificial intelligence: the belief that logic, when structured, could create creativity. Her vision continues to influence how we understand creativity in artificial intelligence today.

## 2.4 World War II and the Rise of Machines

The urgency of war accelerated invention.
In the 1940s, engineers built the first electronic computers — **ENIAC**, **Colossus**, and **Zuse Z3** — massive rooms of vacuum tubes that calculated ballistic trajectories and decrypted enemy codes.

For the first time, humans watched machines **solve problems faster than they could**.
The myth of human-only intelligence began to fade.
Each flicker of a vacuum tube represented a thought in binary: electricity as logic, logic as power.

## 2.5 Alan Turing and the Mind Machine

British mathematician **Alan Turing (1912–1954)** asked a daring question:
*Can machines think?*

Turing's model — the **Turing Machine** — was an abstract device that could simulate any logical process through symbols on a tape.
He proved that reasoning could be mechanized.
His later work on code-breaking at Bletchley Park saved lives and demonstrated that **information** was the new weapon of intelligence.

Turing's legacy established that thinking is not about biology — it is about **processing symbols according to rules**.

## 2.6 The Transistor Revolution

In 1947, scientists at Bell Labs invented the **transistor**, replacing bulky vacuum tubes with tiny, reliable components that could switch millions of times per second.
This single invention shrank computers from warehouse size to desktop size — and ultimately to microchips in the palm of our hands.

Each transistor acts like a microscopic **binary gatekeeper**, deciding whether current should flow (1) or stop (0).
Together, billions of transistors perform trillions of logical operations every second — the heartbeat of every intelligent machine.

## 2.7 The Birth of Software

Hardware made machines possible, but **software made them intelligent**.
Programs written in early languages such as **Fortran**, **COBOL**, and **Assembly** taught computers how to follow sequences of logical instructions.
This separation between hardware and software was revolutionary — much like separating the human brain (structure) from the mind (thought).

The binary foundation became a canvas upon which imagination could be encoded.
A few keystrokes could now create worlds of information.

## 2.8 The Information Age

By the 1960s and 1970s, computing left laboratories and entered classrooms, offices, and homes.
Every digital action — from typing to gaming — was powered by binary decisions executed at incredible speed.

As microprocessors evolved, humanity entered what historians call the **Information Age**, where knowledge itself became a resource.
Binary thinking now governed communication, education, medicine, and art.
Every student using a calculator or computer became part of a global experiment in synthetic reasoning.

## 2.9 Binary as a Mirror of Mind

The birth of digital thinking reshaped how humans understood **themselves**.
If a machine could perform logic, calculate, and even "learn," what then separated human intelligence from artificial?

This philosophical question still drives AI research today.
Many scientists argue that **consciousness** is the next frontier — a dimension of intelligence that cannot yet be reduced to binary.
Others believe that with enough layers, data, and feedback, binary systems may one day emulate awareness itself.

### Classroom Activities

1. **Timeline Challenge:** Create a classroom wall timeline showing the evolution from Leibniz to modern AI chips.
2. **Logic Simulation:** Use paper cards marked 1 and 0 to simulate an AND or OR gate.
3. **Biographical Study:** Research Ada Lovelace or Alan Turing and present how their ideas changed education and technology.
4. **Ethics Debate:** Discuss whether machines that follow logic can ever understand morality.

**This journey reveals how human curiosity transformed into the digital systems that power artificial intelligence today.**

### Chapter Summary

The birth of digital thinking was not a single invention but a gradual evolution shaped by philosophy, mathematics, and engineering. Early thinkers such as Aristotle introduced the foundations of logical reasoning, while George Boole later transformed these ideas into Boolean logic—where all statements could be reduced to true or false, or 1 and 0.

This transformation allowed abstract reasoning to be represented mathematically and eventually physically through electrical systems. Engineers discovered that electrical circuits could mirror logical states, enabling machines to process information using binary signals.

Inventors like Charles Babbage and Ada Lovelace expanded these ideas by designing early computational machines and envisioning their creative potential. During World War II, the development of early computers demonstrated that machines could solve complex problems faster than humans, marking a major shift in how intelligence was understood.

Alan Turing further advanced this concept by showing that reasoning could be mechanized, laying the groundwork for artificial intelligence. The invention of the transistor revolutionized computing by making machines smaller, faster, and more reliable, while the development of software allowed machines to follow structured instructions.

As computing entered everyday life, society transitioned into the Information Age, where digital systems became central to communication, education, and innovation. This chapter highlights that digital thinking is not just about machines—it reflects humanity's ability to model, replicate, and expand its own reasoning processes.

## Reflection Section

### Student Reflection

Respond to the following questions in complete sentences:

How did early philosophers contribute to the development of digital thinking?

Why was Boolean logic important in the creation of computers?

What role did World War II play in advancing computing technology?

How did the invention of the transistor change the future of computing?

In your opinion, what is the most important moment in the history of digital thinking and why?

How does digital thinking reflect human reasoning?

What ethical questions arise when machines begin to "think"?

## 📄 Worksheet: Understanding the Evolution of Digital Thinking

## Part A – Vocabulary Matching

Match each term with the correct description:

A. Boolean Logic
B. Transistor
C. Turing Machine
D. Analytical Engine
E. Information Age

___ A mathematical system based on true/false values

___ A device that controls electrical signals in modern computers

___ A theoretical model that demonstrates how machines can process logic

___ A mechanical design that introduced concepts of memory and processing

___ A period when digital technology became central to society

## Part B – Short Answer

What is the significance of Aristotle's contribution to logic?

How did electrical circuits represent binary logic?

Who was Ada Lovelace, and why is she important?

What did Alan Turing prove about machines and thinking?

Why is the transistor considered a revolutionary invention?

## Part C – Critical Thinking

How did the transition from mechanical machines to electronic computers change society?

Do you think machines can truly think, or are they only following instructions? Explain your reasoning.

## ✅ Answer Key

### Worksheet Answers

#### Part A

A

B

C

D

E

**Part B (Sample Answers)**
6. Aristotle introduced logical reasoning through structured arguments
7. Circuits represented 1 as "on" and 0 as "off"
8. Ada Lovelace envisioned machines creating patterns, music, and more

9. Turing showed that machines could simulate logical thinking
10. It made computers smaller, faster, and more efficient

**Part C (Sample Answers)**
11. It made technology faster, more accessible, and widely used
12. Answers will vary (encourage reasoning and justification)

### Quiz Answers

C

C

B

C

B

**Short Answer (Sample Responses)**
6. Binary connects math (logic) with electricity (on/off signals)
7. He proved machines could simulate reasoning and laid the foundation for AI

## Chapter 3 – Logic Gates: The Neurons of Machines

## Overview

If binary digits are the *letters* of the digital alphabet, **logic gates** are its *grammar*.
They determine how 1s and 0s combine, interact, and transform into higher forms of reasoning.
Logic gates allow machines to "decide," "compare," and "respond" — the same way neurons in a brain transmit signals to create thought.
In this chapter, we explore how microscopic switches evolve into the decision-making circuits that form the **neural core of intelligent systems**.

### Learning Objectives

By the end of this chapter, students will be able to:

1. Define a logic gate and explain its role in digital computation.
2. Describe the basic types of gates — AND, OR, NOT, NAND, NOR, XOR, XNOR — and their truth tables.
3. Compare logic-gate function to biological neurons.
4. Explain how multiple gates form complex circuits and microprocessors.

5. Recognize how logic gates underpin AI decision-making and neural networks.

## 3.1 The Essence of Logic

Every digital process reduces to a single question:

Should the current pass (1) or stop (0)?

Logic gates are tiny decision points that answer that question billions of times each second. They are physical components—built from **transistors**, diodes, or photonic switches—that translate **Boolean logic** into **electrical action**.
When chained together, they create pathways that can store memory, process data, and mimic reasoning.

## 3.2 The Basic Gates

#### AND Gate

Output = 1 only if *both* inputs are 1.
It mirrors the statement "A and B must both be true."

| A | B | Output |
|---|---|---|
| 0 | 0 | 0 |
| 0 | 1 | 0 |
| 1 | 0 | 0 |
| 1 | 1 | 1 |

### OR Gate

Output = 1 if *either* input is 1.
This reflects inclusive reasoning: "A or B (or both)."

### NOT Gate

Also called an *inverter*; it flips the input—turning 1 into 0 and 0 into 1.

### NAND and NOR Gates

These are combinations of the above:

- **NAND** = NOT (AND) → Output is 0 only if both inputs are 1.
- **NOR** = NOT (OR) → Output is 1 only if both inputs are 0.

### XOR and XNOR Gates

- **XOR** (Exclusive OR) = 1 if exactly one input is 1.
- **XNOR** = the opposite — 1 if both inputs are the same.

Together, these seven gate types form the **alphabet of reasoning** used in every processor, AI chip, and robotic brain.

## 3.3 From Gate to Circuit

A single logic gate performs a simple operation, but intelligence arises when **millions of them** are connected.
When arranged in patterns, they form:

- **Adders** (for arithmetic)
- **Comparators** (for evaluation)
- **Flip-flops** (for memory)
- **Multiplexers** (for decision routing)

In essence, a circuit is a **logical sentence** written in binary grammar.
Each path, like a neuron's axon, carries a signal that can branch, merge, and influence other decisions.

## 3.4 Neurons and Logic Gates

A **neuron** receives inputs through dendrites, sums them, and fires if the total exceeds a threshold.
A **logic gate** does the same with electrical inputs.

| Biological Neural Path | Digital Equivalent |
|---|---|
| Dendrites (Inputs) | Input wires (A,B) |
| Axon (Output) | Output wire |
| Synaptic weight | Electrical voltage or logic state |
| Firing threshold | Gate condition |

Both systems reduce to the same principle: **signal integration and activation.**
This is why early AI researchers described neural networks as "electronic brains."
They are not literal brains—but the logic is astonishingly similar.

## 3.5 Combinational and Sequential Logic

- **Combinational Logic**: Output depends only on current inputs (e.g., arithmetic circuits).
- **Sequential Logic**: Output depends on current inputs *and previous states* (e.g., memory, counters).

Sequential logic introduces *time* into the equation.
When a circuit can remember its past, it crosses the threshold from reaction to *computation*—the earliest step toward artificial memory.

## 3.6 The Transistor as a Switch

At the heart of every logic gate lies the **transistor**—a nanoscale device that controls electrical flow.
Think of it as a valve for electrons.
When voltage is applied to its gate terminal, current flows; when removed, current stops.
Billions of these tiny switches flicker in harmony, generating the rhythmic heartbeat of a microprocessor.
Every "thought" a computer has is really a symphony of **on/off** pulses orchestrated by transistors.

## 3.7 From Logic to Learning

How does a set of rigid logic gates create flexible intelligence?
The secret lies in **layering** and **feedback**.
When outputs of one gate network feed into others, complex patterns emerge—much like neural pathways in the brain strengthen through repetition.

Modern AI chips, such as GPUs and TPUs, contain specialized logic optimized for matrix math and neural activation.
Each operation still depends on binary rules, yet together they form the **adaptive networks** that power vision, speech, and decision-making systems.

## 3.8 Binary Precision and Error Tolerance

Because binary gates operate in absolutes—on/off—they resist ambiguity and noise.
This is why digital signals maintain accuracy over long distances.
However, that same rigidity can make systems fragile when dealing with real-world uncertainty.
AI research introduces probabilistic layers above the binary base to simulate **gray areas**, enabling machines to reason with partial truths instead of strict 1s and 0s.

## In-Class Activities

1. **Build a Gate Model** – Use paper circuits or simulation software to demonstrate AND and OR operations.
2. **Neuron vs Gate Comparison** – Draw parallel diagrams of a neuron and a logic gate; label their shared features.
3. **Truth Table Lab** – Create truth tables for all seven basic gates.
4. **Circuit Challenge** – Design a mini-alarm system using only logic gates (e.g., motion AND light sensor = alarm ON).

# 📘 Chapter 3 – Logic Gates: The Neurons of Machines

## Chapter Summary

Logic gates are the fundamental building blocks of digital computation. While binary digits (1s and 0s) represent information, logic gates determine how that information is processed, combined, and transformed into meaningful outputs.

Each logic gate performs a specific operation based on Boolean logic. Basic gates such as AND, OR, and NOT allow machines to make simple decisions, while more advanced gates like NAND, NOR, XOR, and XNOR expand these capabilities. These gates function as decision points, controlling whether electrical signals pass through a circuit.

When multiple logic gates are connected, they form circuits capable of performing complex tasks such as calculations, comparisons, and memory storage. These circuits are the foundation of microprocessors and modern computing systems.

Logic gates also share similarities with biological neurons. Both systems receive inputs, process signals, and produce outputs based on certain conditions. This parallel has influenced the development of artificial intelligence, where digital systems mimic aspects of human neural processes.

The transistor, a tiny electronic switch, powers every logic gate. Billions of transistors working together allow computers to perform trillions of operations per second.

Although logic gates operate on strict binary rules, modern AI systems build upon this foundation by incorporating layers that allow for flexible and probabilistic reasoning. This combination enables machines to move from simple decision-making to advanced learning and intelligent behavior.

## Reflection Section

### Student Reflection

Respond to the following questions in complete sentences:

How do logic gates help computers make decisions?

What is the difference between basic gates (AND, OR, NOT) and advanced gates (XOR, NAND, etc.)?

How are logic gates similar to neurons in the human brain?

Why are transistors important in digital systems?

How do simple logic gates create complex systems?

What does it mean for a system to move from logic to learning?

Why might binary systems need additional layers to handle real-world problems?

## Worksheet: Understanding Logic Gates

### Part A – Vocabulary Matching

Match each term with the correct definition:

A. AND Gate
B. OR Gate
C. NOT Gate
D. Transistor
E. Circuit

___ Outputs 1 only if both inputs are 1

___ Flips the input from 1 to 0 or 0 to 1

___ A network of connected logic gates

___ Outputs 1 if at least one input is 1

___ A tiny switch that controls electrical flow

### Part B – Short Answer

What is the main function of a logic gate?

What is the output of an AND gate when both inputs are 1?

What is the output of an OR gate when one input is 1?

What does a NOT gate do to an input signal?

Why are circuits important in computing systems?

### Part C – Critical Thinking

Explain how logic gates can be combined to perform complex operations.

Why is it helpful to compare logic gates to neurons?

## Quiz: Chapter 3 Assessment

## Multiple Choice

Logic gates are used to:

A. Store emotions
B. Process and control data
C. Create music
D. Display images

Which gate outputs 1 only when both inputs are 1?
A. OR
B. NOT
C. AND
D. XOR

What does a NOT gate do?
A. Adds numbers
B. Flips the input
C. Stores data
D. Combines signals

What is the role of a transistor?
A. To display images
B. To control electrical flow
C. To store memory
D. To create sound

Circuits are made up of:
A. Only wires
B. Logic gates
C. Screens
D. Batteries

## Short Answer

How do logic gates contribute to artificial intelligence?

Why are logic gates considered the "neurons" of machines?

## Answer Key

## Worksheet Answers

### Part A

A

C

E

B

D

**Part B (Sample Answers)**
6. To process inputs and produce outputs based on logic
7. Output = 1
8. Output = 1
9. It reverses the signal
10. Circuits allow multiple gates to work together to perform complex tasks

**Part C (Sample Answers)**
11. By connecting multiple gates, systems can perform calculations, store memory, and make decisions
12. Both process inputs and produce outputs based on conditions

## Quiz Answers

B

C

B

B

B

**Short Answer (Sample Responses)**
6. They process data and support decision-making in AI systems
7. Because they receive inputs, process signals, and produce outputs like neurons

# Chapter 4 – Encoding the World

## Overview

The digital world we live in — our photos, voices, documents, and music — is built from one universal alphabet: **binary**.
Every image we see, every sound we hear, every word we type, has been **encoded** into streams of 1s and 0s.

This chapter explains how computers translate reality into binary form — how light, sound, and language are broken down, measured, and stored as patterns of energy.
Through encoding, machines learn to **see**, **hear**, and **understand**.

### Learning Objectives

By the end of this chapter, students will be able to:

1. Define encoding and explain its role in data representation.
2. Describe how visual, audio, and text data are converted into binary.
3. Understand how pixels, sampling, and ASCII codes work.
4. Explain the relationship between digital accuracy and data size.
5. Recognize encoding as a bridge between the physical world and machine intelligence.

### 4.1 The Meaning of Encoding

To encode means to **translate** information into a form that another system can understand.
Humans do this with languages — converting thoughts into words, sounds, or symbols.
Computers do it with **binary** — converting everything into 1s and 0s.

Encoding allows a computer to represent the richness of the real world through patterns of electric signals. It transforms physical energy into logical structure — **a bridge from atoms to algorithms**.

### 4.2 Seeing the World: Encoding Images

### Pixels and Patterns

A digital image is not a picture — it's a **map of light values**.
Every photograph is divided into thousands or millions of tiny squares called **pixels**.
Each pixel holds color information stored as binary numbers.

- Black and white images use **grayscale values** (0 for black, 255 for white).
- Color images use **RGB encoding**, which combines red, green, and blue light intensities.

For example:

- Pure red → (255, 0, 0)
- Pure green → (0, 255, 0)
- White → (255, 255, 255)

Each color channel is an 8-bit binary number (0–255), creating **24 bits per pixel** in full color.
A single high-resolution image may contain millions of these binary triplets.

### Compression

Since storing billions of binary digits is inefficient, computers use **compression algorithms** like JPEG or PNG to reduce file size without losing visible detail.
These methods remove redundant information while preserving the image's essence — much like summarizing a story without changing its meaning.

## 4.3 Hearing the World: Encoding Sound

Sound travels as waves of vibration.
To record it, computers perform **sampling** — measuring the wave's height (amplitude) at very small intervals of time.

- Each sample is turned into a binary value representing loudness.
- The faster we sample, the more accurate the digital sound becomes.

For example:

- **44,100 samples per second** (44.1 kHz) = CD-quality audio.
- Each sample is stored as 16 bits (2 bytes).

This means every second of audio equals **88,200 bytes** of data before compression.

Formats like **MP3**, **WAV**, and **FLAC** differ in how they compress this information, balancing between **quality** and **file size**.

## 4.4 Reading the World: Encoding Text

Language itself can be digitized through **character encoding systems**.
The most common standard is **ASCII** (American Standard Code for Information Interchange).

Each character is represented by a unique 8-bit binary number:

| Character | Binary | Decimal |
|---|---|---|
| A | 01000001 | 65 |
| B | 01000010 | 66 |
| a | 01100001 | 97 |
| space | 00100000 | 32 |

When you type the word **HELLO**, your computer reads it as:

```
01001000 01000101 01001100 01001100 01001111
```

Unicode, a newer standard, expanded ASCII to include thousands of symbols and languages — allowing machines to communicate across cultures.

## 4.5 Measuring Accuracy: The Trade-Off

Digital encoding always balances **precision** and **efficiency**.

- **Higher precision** → More bits, larger files, slower transmission.
- **Lower precision** → Fewer bits, smaller files, faster transfer, but less detail.

This trade-off mirrors human perception: we don't always need perfect accuracy to understand meaning.
In AI systems, this concept becomes vital — compression, abstraction, and approximation allow machines to learn patterns instead of memorizing every detail.

## 4.6 Binary Patterns in Motion

Videos combine both **image and sound encoding**.
Each frame of a video is a still image (binary pixels), while audio plays alongside it in synchronized time intervals.

Common formats like **MP4**, **MOV**, and **AVI** use layered encoding systems to store video efficiently.
AI algorithms such as **computer vision** and **object tracking** read these frames to identify movement, gestures, or even facial expressions — all decoded from streams of 1s and 0s.

## 4.7 The Role of Encoding in AI

Encoding is how AI *perceives the world.*
When an AI sees a photo, it doesn't see faces — it sees **numerical matrices** of color intensity.
When it listens, it doesn't hear melody — it interprets **frequency patterns** in data.
Natural Language Processing (NLP) systems convert words into **vector embeddings**, where meaning becomes measurable as numbers.

Thus, encoding is not just data storage; it is **data transformation** — turning sensory information into the raw material for learning and reasoning.

## 4.8 Decoding: Bringing It Back to Life

When we open a photo, listen to a song, or read a file, the process reverses.
The computer **decodes** binary values back into visual light, sound waves, and text.
This back-and-forth exchange — encoding and decoding — forms the **heartbeat of digital communication**.
It's the same process used by AI models that translate languages, generate art, or simulate human conversation.

## In-Class Activities

1. **Image Pixel Lab:** Zoom into a digital image to observe individual pixels and note RGB values.
2. **Sound Sampling Exercise:** Use audio software to visualize sound waves and see how sampling works.
3. **ASCII Challenge:** Write your name in ASCII binary code and decode a partner's name.
4. **Compression Experiment:** Compare the quality of an image at 100%, 50%, and 10% compression levels.

## Chapter 4: Encoding the World

### Summary

Encoding is the process of translating real-world information into a format that computers can understand, primarily using binary (1s and 0s). This process allows machines to represent images, sound, and text as structured data.

Images are encoded using pixels, where each pixel stores color information through RGB values. Sound is encoded through sampling, where sound waves are measured and converted into binary values representing amplitude over time. Text is encoded using systems like ASCII and Unicode, which assign binary numbers to characters.

Encoding always involves a trade-off between accuracy and efficiency. Higher precision requires more data, while lower precision reduces file size but may lose detail. Compression techniques help manage this balance.

In artificial intelligence, encoding plays a critical role. AI systems interpret the world not through human senses but through numerical patterns derived from encoded data. Decoding reverses this process, transforming binary back into human-understandable forms like images, audio, and text.

### Reflection

Why is encoding necessary for computers to understand the real world?

How does encoding transform something physical (like sound or light) into something digital?

What is the trade-off between accuracy and file size, and why does it matter?

How does encoding allow artificial intelligence to "see" or "hear"?

Can you think of a time when lower quality (like a compressed image or audio file) was still acceptable? Why?

## Worksheet

### Section A: Vocabulary Matching

Match the term with its correct description:

A. Pixel
B. Sampling
C. ASCII
D. Compression
E. Binary

___ A system that represents characters using numbers

___ Measuring sound waves at intervals

___ Reducing file size while keeping essential information

___ The smallest unit of a digital image

___ A system of 1s and 0s used by computers

### Section B: Short Answer

What does it mean to encode data?

How are images stored using pixels?

What happens when sound is sampled at a higher rate?

What is the purpose of compression?

How does encoding relate to artificial intelligence?

### Section C: Applied Thinking

If you reduce the number of pixels in an image, what happens to quality and file size?

Why might a music streaming service use compressed audio instead of full-quality audio?

Explain how encoding acts as a "bridge" between the physical and digital world.

## Quiz (10 Questions)

### Multiple Choice

What is encoding?
A. Deleting data
B. Translating data into binary form
C. Compressing files
D. Storing memory

What is a pixel?
A. A sound wave
B. A unit of text
C. A small unit of an image
D. A type of algorithm

What does RGB stand for?
A. Red, Green, Blue
B. Random Grid Binary
C. Real Graphic Base
D. Rapid Green Bit

What is sampling in audio?
A. Deleting sound
B. Measuring sound waves at intervals
C. Increasing volume
D. Compressing audio

What does ASCII represent?
A. Images
B. Sound waves
C. Characters using numbers
D. Video frames

What happens when sampling rate increases?
A. Lower quality
B. Smaller file size
C. Higher accuracy
D. No change

What is compression used for?
A. Increasing file size
B. Reducing file size
C. Adding color
D. Encoding sound

What do AI systems “see” when analyzing images?
A. Faces
B. Objects
C. Numbers and patterns
D. Colors only

What is the main trade-off in encoding?
A. Speed vs power
B. Accuracy vs file size
C. Color vs sound
D. Memory vs CPU

What does decoding do?
A. Converts binary back into usable form
B. Deletes data
C. Compresses files
D. Stores data

## Answer Key

## Worksheet Answers

### Section A:

C

B

D

A

E

### Section B (Sample Answers):

Encoding means converting information into a format (binary) that computers can understand.

Images are stored as pixels, each containing color values represented in binary.

Higher sampling rates produce more accurate sound but increase file size.

Compression reduces file size while maintaining acceptable quality.

Encoding allows AI to process real-world data as numerical patterns.

**Section C (Sample Answers):**

Quality decreases, but file size becomes smaller.

To save bandwidth and allow faster streaming.

It converts physical signals (light, sound) into digital data (binary).

### Quiz Answers

B

C

A

B

C

C

B

C

B

A

# Chapter 5 – The Binary Brain

## Overview

Every intelligent system, whether biological or artificial, needs a brain — a center that processes information, makes decisions, and learns from experience.
In computers, that "brain" is built from **billions of binary operations** happening every second. Although it lacks emotion or awareness, the binary brain mirrors human cognition in surprising ways: it stores memory, processes logic, detects patterns, and adapts to new data.

In this chapter, we explore how the binary brain came to exist, how it works, and how it continues to evolve toward higher forms of machine intelligence.

## Learning Objectives

By the end of this chapter, students will be able to:

1. Explain how binary circuits and processors mimic biological brain functions.
2. Describe how data flows through computer architectures.
3. Identify the main components of the computer's "brain."
4. Understand the relationship between memory, processing, and learning.
5. Reflect on how binary systems form the cognitive backbone of artificial intelligence.

## 5.1 The Architecture of Thought

The human brain consists of about **86 billion neurons** that transmit electrical signals through synapses.
The computer's binary brain, though made of silicon instead of cells, performs a similar role through **transistors and logic gates**.

Both systems rely on:

- **Electrical impulses** for communication.
- **Networks of connections** for pattern recognition.
- **Feedback loops** for learning and adaptation.

While the biological brain evolved through nature, the binary brain was **engineered through logic** — an artificial reflection of natural intelligence.

Computers do not think like humans—they follow instructions very quickly."

## 5.2 The CPU: Core Processing Unit

At the center of every digital brain lies the **CPU (Central Processing Unit)** — the "thinking core" that executes binary instructions.
Each instruction is a microscopic command, such as:

- Move data from one location to another.
- Add or compare two values.
- Jump to another operation depending on a condition.

Billions of these instructions happen per second, measured in **GHz (gigahertz)**.
Each tick of the processor's clock synchronizes millions of binary gates, producing the rhythm of machine thought.

This is where **binary meets time** — each clock pulse represents a decision made at light speed.

## 5.3 Memory: The Digital Mind

Just as the human brain separates short-term and long-term memory, computers store data in multiple layers:

- **Registers** – ultra-fast memory built into the CPU.
- **Cache** – quick-access storage for frequently used data.
- **RAM (Random Access Memory)** – temporary workspace for active tasks.
- **Hard Drive / SSD** – long-term memory for files and programs.

Every piece of data, from a student essay to a scientific model, is broken into binary blocks and stored for retrieval.
Memory allows machines to remember **past states**, which is essential for learning and reasoning.

## 5.4 Neural Parallels

A biological neuron processes information by summing inputs and deciding whether to fire a signal.
A **digital processor** does something similar — summing input voltages, comparing them against a threshold, and sending output if conditions are met.

| Human Brain | Binary Brain |
|---|---|
| Neurons | Transistors |
| Synapses | Logic gates and buses |
| Spikes | Clock pulses |
| Neurotransmitters | Electrical signals |
| Plasticity | Machine learning feedback |

Although their materials differ, both systems are **information processors**.
They transform signals into structure — energy into meaning.

## 5.5 The Binary Pathway: From Input to Output

Every action a computer performs follows the **binary loop**:

1. **Input:** Sensors, keyboards, or cameras provide raw data.
2. **Processing:** CPU interprets the binary instructions.
3. **Storage:** Data temporarily resides in memory.
4. **Output:** Results appear on screen, speaker, or network.
5. **Feedback:** The system adjusts based on new data or errors.

This flow mirrors how the human brain perceives, reasons, and reacts — demonstrating how binary can model cognition itself.

## 5.6 Machine Perception

The binary brain learns to "see" through **image recognition**, to "hear" through **audio processing**, and to "understand" through **language models**.
Each sensory input becomes a pattern of binary numbers — the brain's raw material.
AI systems use **neural networks** built on binary foundations to detect faces, interpret emotions, and even predict human intentions.

Though mechanical, the binary brain processes perception with mathematical precision, proving that **logic can emulate awareness** at a structural level.

## 5.7 Parallel Processing: Thinking in Many Streams

Unlike early computers that processed one task at a time, modern systems employ **parallel processing** — many binary circuits working together simultaneously.

This mirrors the human brain's **distributed cognition**, where different regions handle vision, language, and movement at once.
In computing, this is achieved through:

- **Multi-core CPUs**
- **GPUs (Graphics Processing Units)**
- **Neural Processing Units (NPUs)**

Each unit processes thousands of binary streams at once, allowing modern AI to train massive models — the binary equivalent of thinking in parallel.

## 5.8 The Binary Brain and Artificial Learning

When an AI model learns, it adjusts millions of numerical weights — each change represented as binary code.
These binary shifts accumulate to form new patterns of recognition and decision-making.

Learning algorithms like **backpropagation** (in neural networks) depend on simple binary comparisons:

Was the prediction correct (1) or not (0)?

Even complex systems like **Deep Learning** rest on this binary foundation.
The machine's "brain" doesn't *know* — it *calculates*.
Yet through repetition, it begins to approximate understanding.

## 5.9 The Future of the Binary Brain

The binary brain continues to evolve beyond its silicon origins.
Emerging research explores:

- **Neuromorphic chips** – processors that imitate brain architecture.
- **Quantum processors** – where qubits extend beyond 1 and 0 into superposition.
- **Optical computing** – using light instead of electrons for logic.

Even with these advances, binary logic remains the **root layer** — the structural DNA of all machine intelligence.

## In-Class Activities

1. **Brain Comparison Chart:** Create a side-by-side poster comparing the biological brain and the binary brain.
2. **CPU Simulation:** Use a CPU emulator to visualize how instructions are executed.
3. **Binary Memory Map:** Diagram how data moves from input through RAM to output.
4. **Mini AI Lab:** Use a basic neural-network simulator to show how binary weights adjust during learning.

# Chapter 5: The Binary Brain

## Summary

The binary brain is the foundation of all modern computing systems. Just as the human brain processes information through neurons and electrical signals, computers process data through transistors, logic gates, and binary operations. At the center of this system is the CPU, which executes billions of instructions per second, forming the core of machine decision-making.

Memory systems—including registers, cache, RAM, and storage—allow computers to store and retrieve information, much like human short-term and long-term memory. Data flows through a continuous loop of input, processing, storage, output, and feedback, mirroring human cognition.

Artificial intelligence builds on this binary foundation. Machines interpret the world through numerical data, enabling them to recognize patterns, process language, and simulate perception. With advancements like parallel processing, GPUs, and neural networks, machines can handle multiple tasks simultaneously, increasing their ability to learn and adapt.

Although emerging technologies such as quantum computing and neuromorphic chips are expanding possibilities, binary logic remains the core structure underlying all machine intelligence.

## Reflection

In what ways is the computer's binary brain similar to the human brain?

How does the CPU act as the "thinking center" of a computer?

Why is memory important for both humans and machines?

How does the input–process–output cycle compare to human thinking and decision-making?

Do you think machines can ever truly "understand," or are they only calculating? Explain your reasoning.

## Worksheet

## Section A: Vocabulary Matching

Match each term with its correct description:

A. CPU
B. RAM
C. Transistor
D. Parallel Processing
E. Logic Gate

___ The main processing unit of a computer

___ A temporary workspace for active data

___ A tiny electronic switch that processes binary signals

___ Performing multiple operations at the same time

___ A component that performs basic logical operations

## Section B: Short Answer

What is the role of the CPU in a computer system?

How is computer memory similar to human memory?

What is the difference between RAM and long-term storage?

What is parallel processing, and why is it important?

How does AI use the binary brain to learn?

### Section C: Applied Thinking

If a computer had no memory, what problems would it face?

Why are GPUs important for artificial intelligence?

Explain how machines "learn" even though they do not think like humans.

## Quiz (10 Questions)

### Multiple Choice

What is the "brain" of a computer?
A. RAM
B. CPU
C. Hard drive
D. Monitor

What are transistors used for?
A. Storing images
B. Processing binary signals
C. Playing sound
D. Displaying data

What type of memory is temporary?
A. Hard drive
B. SSD
C. RAM
D. Cloud storage

What does the CPU execute?
A. Images
B. Binary instructions
C. Sound waves
D. Colors

What is parallel processing?
A. One task at a time
B. Multiple tasks at the same time
C. Storing data
D. Compressing files

What is similar between neurons and transistors?
A. Both store images
B. Both process signals

C. Both create sound
D. Both display text

What is the purpose of memory in a computer?
A. To delete data
B. To store and retrieve information
C. To increase brightness
D. To compress files

What do AI systems rely on to learn?
A. Emotions
B. Binary calculations and data patterns
C. Human thinking
D. Imagination

What is a GPU used for?
A. Printing documents
B. Handling graphics and parallel processing
C. Typing text
D. Storing files

What is the future of the binary brain exploring?
A. Paper storage
B. Mechanical systems
C. Quantum and neuromorphic computing
D. Typewriters

## Answer Key

## Worksheet Answers

### Section A:

A

B

C

D

E

**Section B (Sample Answers):**

The CPU processes instructions and performs calculations.

Both store and retrieve information for use.

RAM is temporary, while storage is long-term.

Parallel processing allows multiple tasks to be completed at once, improving speed and efficiency.

AI learns by adjusting numerical values (binary weights) based on data and feedback.

**Section C (Sample Answers):**

It could not store information or complete tasks properly.

GPUs handle many calculations at once, making them ideal for AI training.

Machines learn by adjusting data patterns and improving predictions over time.

## Quiz Answers

B

B

C

B

B

B

B

B

B

C

# Chapter 6 – Quantum Leaps Beyond Binary

## Overview

For more than half a century, digital intelligence has lived in a world of *1s and 0s*.
Every app, robot, and algorithm we use depends on that simple duality — **on or off**, **true or false**, **yes or no**.
But nature doesn't think in absolutes.

At the smallest scales of reality, particles exist in **many states at once**.
This discovery gave birth to a new frontier of computation — **quantum computing** — where bits evolve into **qubits**, and logic expands beyond the limits of binary.
In this chapter, we explore how quantum principles reshape our understanding of computation, data, and the future of artificial intelligence.

### Learning Objectives

By the end of this chapter, students will be able to:

1. Explain the difference between classical bits and quantum bits (qubits).
2. Understand the principles of superposition and entanglement.
3. Describe how quantum computing extends beyond binary logic.
4. Identify potential applications of quantum AI.
5. Reflect on how quantum thinking challenges our current definitions of intelligence.

### 6.1 From Binary to Quantum

Classical computing depends on **definitive states** — a bit can be either 1 or 0, never both.
Quantum computing, by contrast, operates on **probabilities**.
A qubit can exist as 1, 0, or any combination of both simultaneously.

This ability comes from the strange physics of quantum mechanics — a world where particles can spin in multiple directions, exist in multiple locations, and even influence each other instantly over distance.

Quantum computing doesn't abandon binary; it **elevates it**.
Where binary logic is linear, quantum logic is **dimensional**.

Quantum computing is still developing and is not yet widely used in everyday devices."

### 6.2 The Qubit: A Quantum Bit

A **qubit** is the basic unit of quantum information.
It is typically represented by the quantum state of a physical particle — such as an electron, photon, or superconducting circuit.

Unlike a bit, which must be 0 *or* 1, a qubit exists in a **superposition** — both 0 and 1 until measured.
This allows quantum computers to process **many possibilities at once**.

Mathematically, a qubit's state can be written as:

$$|\psi\rangle = \alpha\,|0\rangle + \beta\,|1\rangle$$

where $\alpha$and $\beta$are complex numbers representing probabilities.
When measured, the qubit collapses into either 0 or 1, but before that moment, it holds *all potential answers simultaneously*.

## 6.3 Superposition: Thinking in All Directions

Superposition allows a quantum computer to explore multiple outcomes at once.
Imagine flipping a coin that is **heads and tails at the same time** until you look at it.

While a classical computer tests every combination sequentially, a quantum computer evaluates **many paths simultaneously**, making it exponentially faster for certain problems — such as factoring large numbers, optimizing routes, or simulating molecules.

Superposition transforms computation from a flat line into a **sphere of possibilities**.

## 6.4 Entanglement: The Quantum Connection

Entanglement is one of the most mysterious features of quantum systems.
When two qubits become **entangled**, the state of one instantly affects the other — even if they are light-years apart.

This phenomenon allows for **instant coordination** between quantum bits, forming the basis of quantum communication and parallel computation.
In the language of intelligence, entanglement represents **collaboration without delay** — a natural model for connected thinking.

## 6.5 Quantum Logic Gates

Just like classical computers use logic gates (AND, OR, NOT), quantum computers use **quantum gates** to manipulate qubits.

However, these gates rotate qubits in multidimensional space rather than switching them simply on or off.

Examples include:

- **Hadamard Gate (H):** Places a qubit into superposition.
- **Pauli-X Gate:** Flips a qubit (like a NOT gate).
- **CNOT Gate:** Entangles two qubits together.

Each gate performs a reversible transformation, meaning no information is lost — a key principle of quantum coherence.

## 6.6 Quantum Parallelism

Quantum computers can evaluate multiple possibilities at once through **quantum parallelism**. For example, when searching a database, a classical computer must check each entry one by one. A quantum computer, using **Grover's algorithm**, checks all entries simultaneously — dramatically reducing computation time.

This is why quantum systems are expected to revolutionize fields like:

- **Cryptography** (breaking encryption faster)
- **Medicine** (simulating molecules and drug design)
- **AI and Optimization** (learning from massive data spaces)

## 6.7 Quantum AI: A New Kind of Mind

Quantum computing introduces a new form of artificial intelligence — **Quantum AI** — which combines machine learning algorithms with quantum computation.

Quantum AI could:

- Process enormous datasets using parallel superposition.
- Train models faster through quantum-enhanced pattern recognition.
- Solve problems too complex for binary systems, such as global climate simulations or advanced robotics coordination.

These systems would no longer "think" in sequences of 1s and 0s, but in **clouds of probability** — a step closer to how human intuition operates.

## 6.8 The Binary–Quantum Partnership

Quantum computing does not replace binary; it **depends on it**.
Every qubit's measured state ultimately resolves into a 1 or 0 — returning to the binary domain

for interpretation.
In this sense, quantum and binary computing form a **continuum**:

- Binary represents structure and precision.
- Quantum introduces flexibility and probability.

Together, they form a hybrid architecture — the **next evolution of computational intelligence**.

### 6.9 The Future Beyond Binary

Quantum technology is still in its infancy, facing challenges such as **decoherence**, **error correction**, and **temperature control**.
Yet the vision is clear: a world where quantum and classical processors collaborate seamlessly.

Tomorrow's AI systems may no longer "calculate" as we know it — they will **sense**, **correlate**, and **predict** reality through quantum-informed cognition.
This is not the end of binary, but its **transformation into multidimensional reasoning**.

### In-Class Activities

1. **Quantum Coin Flip:** Use coins to demonstrate superposition — heads (1), tails (0), edge (both).
2. **Qubit Simulation:** Explore an online quantum simulator to visualize qubit rotation and entanglement.
3. **Binary vs. Quantum Comparison:** Create a chart showing classical logic gates vs. quantum gates.
4. **Future Vision Essay:** Have students write how quantum AI could change medicine, art, or education.

### Chapter 6 – Quantum Leaps Beyond Binary

## Chapter Summary

Quantum computing represents a major advancement beyond traditional binary systems. While classical computers rely on bits that exist as either 0 or 1, quantum computers use qubits, which can exist in multiple states at once through a principle called superposition.

This ability allows quantum systems to process many possibilities simultaneously, making them potentially far more powerful for certain complex problems. Another key concept is entanglement, where qubits become linked so that the state of one instantly influences another, enabling highly coordinated computation.

Quantum logic gates operate differently from classical logic gates. Instead of simply turning signals on or off, they manipulate qubits in multidimensional space, allowing for reversible and highly efficient transformations.

Quantum parallelism enables computers to evaluate multiple solutions at the same time, significantly improving performance in areas such as cryptography, optimization, and scientific simulations.

Quantum AI combines these capabilities with machine learning, allowing systems to process vast datasets and solve problems beyond the reach of classical computing.

Despite these advancements, quantum computing still relies on binary systems for interpreting results. Together, classical and quantum computing form a hybrid model—combining precision with probability.

Although still developing, quantum technology has the potential to reshape artificial intelligence, computation, and our understanding of information itself.

## Reflection Section

### Student Reflection

Respond to the following questions in complete sentences:

What is the main difference between a bit and a qubit?

What does superposition mean in quantum computing?

How does entanglement change the way information is processed?

Why are quantum computers faster for certain problems?

What is quantum parallelism?

How might quantum AI impact the future of technology?

Why do quantum systems still rely on binary systems?

## Worksheet: Understanding Quantum Computing

## Part A – Vocabulary Matching

Match each term with the correct definition:

A. Qubit
B. Superposition
C. Entanglement
D. Quantum Gate
E. Parallelism

___ A quantum unit of information

___ The ability to exist in multiple states at once

___ When two particles are instantly connected

___ A transformation applied to qubits

___ Processing multiple possibilities at the same time

## Part B – Short Answer

What is the difference between classical and quantum computing?

What happens when a qubit is measured?

Why is superposition important?

What is the role of quantum gates?

How does quantum computing improve problem-solving?

## Part C – Critical Thinking

Explain how quantum computing changes the way we think about information.

Why might quantum AI be more powerful than traditional AI?

## Quiz: Chapter 6 Assessment

## Multiple Choice

A qubit differs from a bit because it:
A. Stores more data
B. Exists in multiple states
C. Uses electricity
D. Is faster

Superposition allows a system to:
A. Store files
B. Run programs
C. Exist in multiple states simultaneously
D. Turn off

Entanglement means:
A. Data is deleted
B. Two qubits are connected
C. Files are stored
D. Signals are lost

Quantum gates:
A. Store memory
B. Control electricity
C. Manipulate qubits
D. Display output

Quantum computing is useful for:
A. Only simple tasks
B. Complex problem solving
C. Printing documents
D. Typing text

## Short Answer

How does quantum computing improve AI systems?

What is the relationship between binary and quantum computing?

## Answer Key

## Worksheet Answers

### Part A

A

B

C

D

E

**Part B (Sample Answers)**
6. Classical uses bits (0 or 1), quantum uses qubits (multiple states)
7. It collapses into either 0 or 1
8. It allows multiple possibilities to be processed at once
9. To manipulate and control qubit states
10. By evaluating many solutions simultaneously

**Part C (Sample Answers)**
11. It introduces probability instead of fixed states
12. It can process larger datasets and solve more complex problems

## Quiz Answers

B

C

B

C

B

**Short Answer (Sample Responses)**
6. It allows faster processing and better pattern recognition
7. Quantum extends binary but still depends on it for final output

# Chapter 7 – The Binary DNA of Algorithms

## Overview

Every living organism is guided by its **DNA** — a code that instructs how to grow, adapt, and survive.
In the same way, every intelligent machine follows its own genetic code: the **algorithm**.

Algorithms are the **binary DNA** of artificial intelligence. They define how machines think, how they solve problems, and how they evolve.
This chapter explores how algorithms are designed, structured, and executed — showing that all computational intelligence, no matter how advanced, begins as a sequence of logical instructions written in binary.

### Learning Objectives

By the end of this chapter, students will be able to:

1. Define what an algorithm is and explain its role in computing.
2. Describe how algorithms are encoded in binary and executed by hardware.
3. Identify the key types of algorithms used in artificial intelligence.
4. Understand how efficiency, complexity, and optimization affect performance.
5. Compare biological DNA and digital algorithms as systems of coded intelligence.

## 7.1 What Is an Algorithm?

An **algorithm** is a step-by-step set of instructions for solving a problem.
It tells a machine *what to do, how to do it, and when to stop*.

In everyday life, recipes, driving directions, or math formulas are all algorithms — sequences of logic that lead to predictable outcomes.
In computers, these instructions are translated into **binary commands**, making them executable by processors.

Every algorithm follows a universal pattern:

1. **Input** – Receive data.
2. **Process** – Perform computations or decisions.
3. **Output** – Produce a result.

This simple cycle forms the backbone of both software and artificial intelligence.

"An algorithm is like a recipe that tells a computer what steps to follow."

## 7.2 Binary Encoding of Instructions

At the hardware level, algorithms exist only as **machine code** — binary sequences like:

```
10110000 01100001
10110001 01100010
00000000
```

Each pattern corresponds to a specific operation:

- Move data to a register.
- Add or compare values.
- Jump to another line of code if a condition is met.

High-level programming languages (like Python or C++) are human-readable versions of these binary commands.
When compiled, they are converted back into the 1s and 0s that the **CPU** understands.

Thus, all intelligence — no matter how elegant it appears — begins as **electrical signals following binary pathways.**

## 7.3 The Algorithm as Digital DNA

Just as biological DNA contains instructions for building a living organism, an algorithm contains instructions for building **digital behavior**.

- DNA uses **nucleotides (A, T, C, G)** to store information.
- Computers use **bits (1s and 0s)**.

Both systems share similar properties:

| Property | Biological DNA | Algorithmic DNA |
|---|---|---|
| Storage Unit | Nucleotides | Bits |
| Transmission | Genetic replication | Data copying |
| Mutation | Genetic variation | Code modification |
| Adaptation | Natural evolution | Machine learning |

This comparison reveals a deep truth: both life and intelligence evolve from **coded patterns** that adapt to environments.

## 7.4 The Structure of Algorithms

Algorithms are often visualized using **flowcharts** or **pseudocode**.
A typical algorithm includes:

1. **Variables** – containers for data.
2. **Control Structures** – logic (IF, THEN, ELSE).
3. **Loops** – repetition (FOR, WHILE).
4. **Functions** – reusable procedures.

Every part of an algorithm operates through **binary decision-making**:

Is the condition true (1) or false (0)?

That single binary question is what drives every software system in the world — from calculators to neural networks.

## 7.5 Algorithmic Efficiency

Not all algorithms are equal. Some solve problems faster and with fewer resources.
Efficiency is measured by **time complexity** and **space complexity** — how much time and memory an algorithm requires.

- **O(1)** – Constant time (fastest).
- **O(n)** – Linear time (depends on input size).
- **$O(n^2)$** – Quadratic time (slower for large inputs).

Understanding efficiency helps AI engineers build systems that can handle massive datasets without slowing down — a crucial factor in deep learning and robotics.

## 7.6 Learning Algorithms in AI

AI uses specialized algorithms that **learn from data** rather than following fixed rules.
Some major types include:

- **Supervised Learning:** Learns from labeled examples (e.g., recognizing cats in photos).
- **Unsupervised Learning:** Finds hidden patterns without labels.
- **Reinforcement Learning:** Learns by trial and error through rewards and penalties.

In all cases, the underlying computations — additions, comparisons, thresholds — are **binary operations** executed billions of times per second.

## 7.7 Evolutionary Algorithms

Inspired by nature, **evolutionary algorithms** simulate the process of natural selection. They create multiple solutions, test them, keep the best ones, and mutate the rest. Over time, the system "evolves" an optimal solution.

This approach mirrors biological evolution, where **genetic variation** drives survival and improvement.
Each generation of code becomes smarter — an algorithm rewriting its own binary DNA.

## 7.8 Algorithmic Intelligence

When many algorithms interact — sorting data, detecting patterns, making predictions — they form **systems of collective intelligence**.
For example:

- A recommendation engine combines sorting, filtering, and clustering algorithms.
- A self-driving car merges vision, navigation, and reinforcement-learning algorithms.

Each of these components runs on binary logic but together simulate **reasoning**, **memory**, and **decision-making** — the hallmarks of intelligence.

## 7.9 Ethical Algorithms

As algorithms influence human life — from job applications to criminal sentencing — the need for **ethical design** has grown urgent.
Ethical algorithms ensure that data is fair, unbiased, and transparent.

Education in **algorithmic ethics** helps future scientists, teachers, and students understand that intelligence without morality can harm as easily as it helps.
Every algorithm should not only compute efficiently — it should also **act responsibly**.

## In-Class Activities

1. **Algorithm Recipe:** Write a recipe for a simple activity (like making a sandwich) as an algorithm.
2. **Flowchart Design:** Draw a flowchart showing how a computer decides whether a password is correct.
3. **Binary-to-Biological Comparison:** Compare human DNA sequencing to computer code execution.
4. **Algorithm Efficiency Test:** Have students simulate sorting 10 numbers using two different sorting methods and measure time.

# Chapter 7 – The Binary DNA of Algorithms

## Chapter Summary

Algorithms are the foundational instructions that drive all computing systems. Just as DNA provides the instructions for life, algorithms provide the instructions for machines to process information, solve problems, and make decisions.

An algorithm follows a simple structure: input, processing, and output. These steps define how data is received, transformed, and returned as results. At the hardware level, all algorithms are encoded in binary, allowing processors to execute instructions through electrical signals.

High-level programming languages translate human-readable instructions into machine code, but ultimately, all computational processes rely on binary operations.

Algorithms share similarities with biological DNA. Both systems store information, replicate processes, adapt over time, and evolve. In computing, this evolution occurs through optimization, machine learning, and code refinement.

The structure of algorithms includes variables, control structures, loops, and functions. Each decision within an algorithm is based on binary logic—true or false.

Efficiency is a key factor in algorithm design. Time and space complexity determine how well an algorithm performs, especially when handling large datasets.

Artificial intelligence relies on learning algorithms such as supervised, unsupervised, and reinforcement learning. These systems improve through repeated exposure to data, adjusting internal values to produce better outcomes.

Evolutionary algorithms mimic natural selection, allowing systems to improve by testing, selecting, and modifying solutions over time.

As algorithms become more powerful, ethical considerations become increasingly important. Responsible algorithm design ensures fairness, transparency, and accountability in systems that impact human lives.

## Reflection Section

### Student Reflection

Respond to the following questions in complete sentences:

What is an algorithm, and why is it important in computing?

How are algorithms similar to biological DNA?

What are the three main steps of an algorithm?

Why is binary important in executing algorithms?

What is the difference between supervised and unsupervised learning?

Why is algorithm efficiency important?

Why do algorithms need ethical considerations?

# Worksheet: Understanding Algorithms

## Part A – Vocabulary Matching

Match each term with the correct definition:

A. Algorithm
B. Binary
C. Loop
D. Efficiency
E. Machine Learning

___ A step-by-step set of instructions

___ Repeating a set of instructions

___ Measuring how well an algorithm performs

___ A system that learns from data

___ A system using 1s and 0s

## Part B – Short Answer

What are the three main steps of an algorithm?

What happens when code is converted into binary?

What is the purpose of loops in an algorithm?

What is time complexity?

Why is machine learning important in AI?

## Part C – Critical Thinking

Explain how algorithms can "evolve" over time.

Why is it important for algorithms to be fair and unbiased?

## Quiz: Chapter 7 Assessment

## Multiple Choice

An algorithm is:
- A. A type of computer
- B. A step-by-step instruction process
- C. A storage system
- D. A programming language

Binary is made up of:
- A. Letters
- B. Colors
- C. 1s and 0s
- D. Sounds

A loop is used to:
- A. Stop a program
- B. Repeat instructions
- C. Store data
- D. Display output

Machine learning allows systems to:
- A. Store files
- B. Learn from data
- C. Print documents
- D. Turn off

Algorithm efficiency refers to:

A. Design style
B. Speed and resource use
C. Color coding
D. Storage size

## Short Answer

How do algorithms support artificial intelligence?

Why is ethical algorithm design important?

## Answer Key

## Worksheet Answers

### Part A

A

C

D

E

B

**Part B (Sample Answers)**

6. Input → Process → Output
7. It becomes machine-readable instructions
8. To repeat tasks efficiently
9. A measure of how long an algorithm takes to run
10. It allows systems to learn and improve from data

**Part C (Sample Answers)**

11. Through updates, optimization, and learning from data
12. To prevent bias and ensure fairness

## Quiz Answers

B

C

B

B

B

**Short Answer (Sample Responses)**
6. They provide the instructions for processing and learning from data
7. Because algorithms impact real-world decisions and must be fair

# Chapter 8 – Data, Decision, and Duality

## Overview

Every intelligent action begins with a decision — a choice between possibilities.
In the digital world, this decision-making process is built on a simple foundation: **duality**.

Binary logic thrives on opposites: *true or false, on or off, yes or no*.
From these opposites emerges the capacity to reason, learn, and adapt.
This chapter explores how binary duality shapes both **human thought** and **machine intelligence**, demonstrating that decisions are the heartbeat of understanding.

### Learning Objectives

By the end of this chapter, students will be able to:

1. Define duality and explain its role in binary logic.
2. Understand how data influences decision-making in computers.
3. Describe how binary contrast (1 and 0) enables reasoning and learning.
4. Explore how duality exists in natural, human, and artificial systems.
5. Reflect on the philosophical and ethical meaning of "decision" in AI.

### 8.1 The Nature of Duality

Duality means **two complementary forces** that define each other through contrast.
Light exists because there is darkness. Truth has meaning because there is falsehood.
In binary systems, this duality becomes mathematical — the existence of two discrete states, **1 and 0**.

This principle reflects one of the universe's oldest patterns: balance.
From the magnetic poles of Earth to the positive and negative charges in atoms, duality forms the rhythm of creation.
In computing, duality becomes the rhythm of logic — the beat that drives artificial thought.

## 8.2 Binary Contrast: The Foundation of Logic

All reasoning in computers stems from binary comparisons:

Is this condition true (1) or false (0)?

This question determines every action — whether to open a file, send a signal, or perform a calculation.
The decision mechanism lies in **logic gates** and **conditional statements**, which evaluate data and guide the flow of execution.

For example:

```
IF temperature > 80 THEN
   turn_on_fan = 1
ELSE
   turn_on_fan = 0
```

This is a simple binary decision — yet millions of such decisions per second form the foundation of artificial intelligence.

## 8.3 Data: The Fuel for Decision-Making

Data is the information machines rely on to make choices.
Without data, an algorithm cannot learn, adapt, or evolve.

AI systems transform raw data into meaningful insight through steps:

1. **Collection** – Gathering facts, numbers, or sensory inputs.
2. **Processing** – Organizing and cleaning data.
3. **Analysis** – Finding patterns and relationships.
4. **Decision** – Acting on patterns to produce an outcome.

Every stage depends on binary judgments — sorting, comparing, and classifying information into structured meaning.

## 8.4 Duality in Human Reasoning

Humans also think in dualities:

- right and wrong
- safe and dangerous
- success and failure

These pairs help the brain simplify complex realities into manageable categories.
However, humans can also **see shades between opposites** — something AI systems are still learning.
Teaching AI to interpret "gray areas" of emotion, ethics, and context is one of the greatest challenges in modern computing.

## 8.5 The Binary Decision Process

In computers, a decision follows a clear chain:

1. **Input:** Receive data (1s and 0s).
2. **Condition:** Compare against a rule or threshold.
3. **Output:** Choose one of two possible actions.

This system of binary branching allows machines to build complex structures from simple logic.
Just as a tree branches from a trunk into countless leaves, binary decisions expand into **decision trees** — the framework behind many AI models.

## 8.6 Decision Trees and Machine Reasoning

A **decision tree** is an AI model that mirrors human reasoning.
Each node represents a question (yes/no), and each branch represents a possible outcome.

Example:

```
Is it raining?
   Yes → Take umbrella
   No  → Wear sunglasses
```

When thousands of such questions are layered together, machines can make predictions about weather, finance, or health.
At the heart of these trees lies binary evaluation — a continuous dialogue of *if...then...else.*

## 8.7 Duality and Learning

Learning occurs when a system repeatedly evaluates outcomes and adjusts its future decisions. Each correction — whether in a neural network or human mind — strengthens or weakens a connection.
The binary outcome (success or failure) becomes the **feedback signal** that drives improvement.

This is the principle behind **reinforcement learning**, where AI agents learn by receiving rewards (1) or penalties (0).
From these binary experiences, machines develop strategies — a digital form of trial and error.

## 8.8 Beyond Duality: Gray Logic and Probability

Real-world decisions are rarely absolute.
To handle uncertainty, AI systems use **fuzzy logic** and **probabilistic reasoning** — methods that allow degrees between 0 and 1.

Example:

- 0.0 → definitely false
- 0.5 → somewhat true
- 1.0 → completely true

This approach helps AI understand real-world complexity, such as interpreting emotions or predicting human behavior.
It is how machines begin to move from **binary judgment** to **contextual understanding**.

## 8.9 Ethical Dualities

Every decision — whether human or machine — carries ethical weight.
AI systems used in healthcare, hiring, or law must navigate moral dualities such as:

- Fairness vs. efficiency
- Privacy vs. transparency
- Freedom vs. control

Teaching AI to make balanced decisions requires embedding **human ethics into machine logic**, ensuring that algorithms serve people rather than replace them.

Binary precision must be guided by moral wisdom.

## In-Class Activities

1. **Duality Map:** Have students list examples of dualities found in nature, society, and computing.
2. **Decision Tree Project:** Design a simple decision tree for choosing transportation to school based on weather.
3. **Binary Reflection:** Write a paragraph explaining how duality helps humans and machines simplify complexity.
4. **Ethics Discussion:** Debate the question: *Should AI always follow logical reasoning even if it conflicts with empathy?*

# Chapter 8 – Data, Decision, and Duality

## Chapter Summary

Decision-making is at the core of both human intelligence and artificial intelligence. In digital systems, this process is built on binary duality—the concept of opposites such as true/false, yes/no, and 1/0.

Binary contrast allows computers to evaluate conditions and make decisions through logic gates and conditional statements. Every action a computer takes is based on these simple comparisons, repeated millions of times per second.

Data serves as the foundation for decision-making. AI systems collect, process, analyze, and act on data to generate meaningful outcomes. Without data, machines cannot learn or adapt.

Humans also rely on duality to simplify complex thinking, using contrasts such as right/wrong or success/failure. However, unlike machines, humans can interpret nuance and context beyond simple opposites.

In computing, decisions follow a structured process: input, condition, and output. This structure expands into decision trees, where layered binary choices allow machines to perform complex reasoning and predictions.

Learning systems improve through feedback. Reinforcement learning, for example, uses binary outcomes such as reward or penalty to guide future decisions.

To handle real-world complexity, AI systems go beyond strict binary logic by using probabilities and fuzzy logic, allowing decisions to exist between 0 and 1.

As AI becomes more integrated into society, ethical considerations become critical. Decisions made by machines must balance fairness, privacy, and responsibility, ensuring that technology supports human values.

# Reflection Section

## Student Reflection

Respond to the following questions in complete sentences:

What is duality, and how does it relate to binary logic?

How do computers use binary decisions to perform tasks?

Why is data important for decision-making in AI?

How is human reasoning different from machine decision-making?

What is a decision tree, and how does it work?

How does reinforcement learning use feedback?

Why are ethical decisions important in artificial intelligence?

# Worksheet: Understanding Data and Decision-Making

## Part A – Vocabulary Matching

Match each term with the correct definition:

A. Duality
B. Data
C. Decision Tree
D. Reinforcement Learning
E. Binary

___ Information used to make decisions

___ A system based on two opposite states

___ A model that uses branching decisions

___ Learning through rewards and penalties

___ A system using 1s and 0s

## Part B – Short Answer

What is binary contrast?

What are the steps in data processing?

How does a computer make a decision?

What is reinforcement learning?

Why is fuzzy logic important?

## Part C – Critical Thinking

Explain how duality helps simplify complex problems.

Why must AI systems consider ethics when making decisions?

## Quiz: Chapter 8 Assessment

### Multiple Choice

Duality refers to:
A. Multiple systems
B. Two opposing states
C. Large data
D. Storage

Binary logic is based on:
A. Colors
B. Numbers
C. 1s and 0s
D. Words

A decision tree is used to:
A. Store data
B. Make structured decisions

C. Display output
D. Compress files

Reinforcement learning is based on:
A. Storage
B. Rewards and penalties
C. Colors
D. Speed

Fuzzy logic allows:
A. Exact answers only
B. No answers
C. Partial or probabilistic answers
D. Binary only

## Short Answer

How does data influence AI decision-making?

Why is ethical decision-making important in AI systems?

## Answer Key

## Worksheet Answers

### Part A

B

A

C

D

E

**Part B (Sample Answers)**
6. Comparing two opposite states (true/false)
7. Collection → Processing → Analysis → Decision
8. By evaluating conditions and selecting an output
9. Learning through rewards and penalties
10. It allows decisions between true and false

**Part C (Sample Answers)**
11. It reduces complexity into manageable choices
12. Because AI decisions impact real-world outcomes

### Quiz Answers

B

C

B

B

C

**Short Answer (Sample Responses)**
6. Data provides the information needed for decisions
7. To ensure fairness, safety, and responsibility

## Chapter 9 – Binary Ethics: Right, Wrong, and the Space Between

## Overview

In a world ruled by data and decisions, how do we teach machines the difference between **right and wrong**?
Binary logic gives us precision, but ethics requires something more — **context, empathy, and wisdom**.

This chapter explores how binary systems form the backbone of moral decision-making in artificial intelligence and why understanding the *space between* 1 and 0 is essential for the responsible design of intelligent machines.

## Learning Objectives

By the end of this chapter, students will be able to:

1. Define binary ethics and explain its role in artificial intelligence.
2. Understand how AI systems make decisions based on programmed values.
3. Discuss the limitations of binary reasoning in moral and social contexts.
4. Identify ethical challenges in algorithm design and data use.
5. Reflect on the balance between logic, fairness, and human judgment.

## 9.1 The Origins of Moral Duality

Since ancient times, humans have defined morality through duality:

- good vs. evil
- right vs. wrong
- justice vs. injustice

These opposites mirror the binary foundations of reasoning — clear, contrasting states that create clarity.
However, unlike machines, humans add emotion, intuition, and culture to these judgments.

When we teach machines to make ethical choices, we are essentially asking:

Can logic alone define morality?

## 9.2 Binary Morality in Machines

AI systems make ethical decisions using binary conditions, such as:

```
IF (harm < benefit) THEN action = 1
ELSE action = 0
```

This is efficient for mechanical reasoning but limited in moral depth.
For instance, a self-driving car might calculate which action causes fewer injuries — but it cannot **feel** compassion or guilt.

Binary systems follow logic; humans follow values.
To align the two, we must embed ethics into algorithms as structured principles.

"AI does not feel emotions—it applies rules based on data."

## 9.3 The Three Pillars of Ethical AI

### 1. Fairness

AI should make decisions without bias.
Binary precision can unintentionally reproduce human prejudice if training data is unbalanced.

### 2. Transparency

Humans must be able to understand how an AI reaches its decisions.
Binary rules must be visible, explainable, and open to review.

### 3. Accountability

Someone — a developer, organization, or government — must take responsibility for the outcomes of AI decisions.

These principles turn binary logic into **ethical logic**, ensuring that intelligence remains under human guidance.

## 9.4 The Space Between 1 and 0

In ethics, not every situation fits neatly into "yes" or "no."
Some choices are morally complex — the gray areas where empathy and circumstance matter.
For example:

- Is it right to prioritize one life over another?
- Should privacy be sacrificed for safety?
- Can an AI ever understand mercy?

These questions reveal that morality often lives **between** binary states.
Fuzzy logic, probabilistic reasoning, and human oversight help machines interpret that middle ground — the **moral continuum** between extremes.

## 9.5 When Data Meets Dilemma

AI systems learn morality from data — but data reflects society, and society is imperfect.
If the data includes stereotypes or inequalities, the AI will learn them too.
This leads to **algorithmic bias**, where binary logic amplifies existing injustices.

For example:

- Facial recognition systems performing poorly on darker skin tones.
- Hiring algorithms preferring male candidates due to biased historical data.

These examples remind us that **ethical AI begins with ethical data**.

## 9.6 The Trolley Problem in Binary Form

A famous ethical dilemma, the *Trolley Problem*, asks:

If a train is about to hit five people, should you pull a lever to redirect it, killing one person instead?

Humans debate this through emotion, culture, and philosophy.
Machines, however, process it as a binary equation — minimize total harm.
While logical, this approach ignores emotional and social consequences.

True intelligence must blend **binary calculation** with **human compassion**.

## 9.7 Programming Moral Logic

Researchers and engineers are now developing frameworks for **machine ethics**, such as:

- **Value Alignment:** Ensuring AI values match human values.
- **Explainable AI (XAI):** Making machine reasoning transparent.
- **Ethical Governance:** Creating laws and policies for responsible AI.

Each of these efforts translates moral concepts into logical, measurable rules — the next step toward an AI that can reason both *rationally* and *responsibly.*

## 9.8 The Role of Education in Binary Ethics

Schools play a critical role in shaping the next generation of AI designers and digital citizens. By teaching binary ethics early, students learn that technology is not just about **what we can build**, but **what we should build**.

Educators can help students:

- Explore the intersection of technology and morality.
- Discuss ethical implications of data collection.
- Reflect on personal responsibility in a digital world.

When logic and ethics grow together, intelligence becomes wisdom.

## 9.9 The Future of Moral Machines

As AI becomes more integrated into daily life, the question of **ethical awareness** grows urgent. Can a machine ever develop a conscience?
Perhaps not in the human sense — but it can simulate fairness, empathy, and justice if guided by ethical design.

In the future, binary systems may evolve beyond calculation — toward **moral cognition**, where logic and compassion work side by side to create responsible intelligence.

### In-Class Activities

1. **Ethical Coding:** Write a short pseudocode for an AI decision that balances fairness and safety.
2. **Data Reflection:** Discuss how biased data might influence AI ethics.
3. **Debate:** "Can an AI ever be truly moral?"
4. **Moral Tree Exercise:** Create a decision tree for a complex moral question and discuss its limitations.

## Chapter 9 – Binary Ethics: Right, Wrong, and the Space Between

## Chapter Summary

Binary ethics explores how artificial intelligence systems make decisions about right and wrong using logical rules. While computers rely on binary logic—1s and 0s—ethical decision-making requires deeper understanding, including context, fairness, and human values.

AI systems make decisions based on programmed conditions, but these decisions are limited by the data and rules they are given. Unlike humans, machines do not experience emotions, empathy, or moral intuition. This creates challenges when AI is used in real-world situations involving complex ethical dilemmas.

To address these challenges, ethical AI is built on three key principles: fairness, transparency, and accountability. These principles help ensure that AI systems make decisions that are unbiased, understandable, and responsible.

However, many ethical situations cannot be reduced to simple binary choices. Real-world decisions often exist in a "gray area" between right and wrong. AI systems use approaches such as fuzzy logic and probabilistic reasoning to better handle this complexity.

Data plays a critical role in shaping AI behavior. If data is biased, AI systems can reproduce and amplify those biases. This highlights the importance of ethical data collection and system design.

Classic dilemmas, such as the trolley problem, demonstrate the limitations of purely logical decision-making. While machines may choose the most efficient outcome, they may overlook emotional or societal factors.

Education plays a vital role in preparing students to understand and guide ethical AI. By combining technical knowledge with ethical awareness, future generations can ensure that AI systems serve humanity responsibly.

## Reflection Section

### Student Reflection

Respond to the following questions in complete sentences:

What is binary ethics?

Why is binary logic not enough for ethical decision-making?

What are the three pillars of ethical AI?

What is meant by the "space between 1 and 0"?

How can biased data affect AI systems?

Why is the trolley problem important in AI ethics?

What role does education play in ethical AI development?

## Worksheet: Understanding Binary Ethics

## Part A – Vocabulary Matching

Match each term with the correct definition:

A. Fairness
B. Transparency
C. Accountability
D. Bias
E. Fuzzy Logic

___ Responsibility for decisions made by AI

___ Decisions made without unfair advantage

___ Openness in how AI systems work

___ Prejudice in data or decision-making

___ Reasoning with degrees between true and false

## Part B – Short Answer

What is binary morality in machines?

Why can AI systems struggle with ethical decisions?

What is algorithmic bias?

What is the purpose of ethical AI principles?

Why are gray areas important in decision-making?

## Part C – Critical Thinking

Explain why AI should not rely only on logic when making decisions.

How can humans ensure that AI systems act responsibly?

## Quiz: Chapter 9 Assessment

## Multiple Choice

Binary ethics refers to:
A. Storage systems
B. Ethical decision-making in AI
C. Programming languages
D. Hardware

Which is NOT a pillar of ethical AI?
A. Fairness
B. Transparency
C. Accountability
D. Speed

Bias in AI comes from:
A. Hardware failure
B. Random errors
C. Biased data
D. Storage issues

Fuzzy logic allows:
A. Only true or false
B. No decisions

C. Degrees between true and false
D. Faster processing

The trolley problem is used to:
A. Build software
B. Test hardware
C. Explore ethical decisions
D. Store data

### Short Answer

Why is fairness important in AI systems?

How can education improve ethical AI?

## Answer Key

### Worksheet Answers

#### Part A

C

A

B

D

E

**Part B (Sample Answers)**
6. AI making decisions using binary logic
7. Because they lack emotion and context
8. When AI reflects unfair patterns from data
9. To guide responsible and fair decision-making
10. Because not all decisions are absolute

**Part C (Sample Answers)**
11. Because real-world decisions require context and empathy
12. By designing systems with ethical principles and oversight

**Quiz Answers**

B

D

C

C

C

**Short Answer (Sample Responses)**
6. To ensure decisions are fair and unbiased
7. By teaching responsibility, ethics, and critical thinking

# Chapter 10 – Binary Learning Systems in Education

## Overview

The 21st-century classroom is no longer defined by chalkboards and textbooks alone — it is powered by **data, logic, and algorithms**.
As technology reshapes every career and community, understanding **binary learning systems** has become as essential as reading or mathematics.

This chapter explores how binary principles — decision-making, logic, structure, and feedback — can transform education into a more interactive, analytical, and personalized experience.
By connecting binary thought to human creativity, schools can prepare students not only to use technology, but to understand **how intelligence itself works.**

### Learning Objectives

By the end of this chapter, students will be able to:

1. Define what a binary learning system is and how it applies to education.
2. Explain how binary logic enhances problem-solving and computational thinking.
3. Identify real-world classroom applications of binary-based models.
4. Understand how AI-driven systems personalize learning through binary data.
5. Reflect on the ethical responsibilities of using intelligent technology in schools.

## 10.1 What Is a Binary Learning System?

A **binary learning system** is an educational framework built around clear logic, measurable feedback, and adaptive responses — much like how a computer processes 1s and 0s.
In this model, each learning decision (correct/incorrect, understood/not understood) becomes data that helps refine the next step.

Binary systems in education operate on three core principles:

1. **Clarity:** Define learning outcomes with precision.
2. **Feedback:** Evaluate performance in real time.
3. **Adaptation:** Adjust teaching strategies based on data patterns.

Just as a computer learns through input and feedback loops, students learn best when information is structured, interactive, and responsive.

## 10.2 The Parallel Between Computers and Classrooms

| Computer System | Learning System |
|---|---|
| Input | Student receives lesson or activity |
| Processing | Student applies logic and understanding |
| Output | Student produces an answer or project |
| Feedback | Teacher or AI system evaluates result |
| Adjustment | Student and teacher adapt strategy |

This mirrored structure shows that binary learning is not just digital — it is **cognitive**.
Every time a student reflects, corrects, and improves, they are performing a natural binary loop: testing hypotheses, measuring outcomes, and recalibrating understanding.

## 10.3 Binary Thinking in the Curriculum

Binary concepts can be taught across multiple subjects:

- **Mathematics:** Boolean logic, probability, and problem decomposition.
- **Science:** Understanding cause and effect, conditional relationships, and feedback mechanisms.
- **Language Arts:** Analyzing argument logic — true/false claims, consistency, and reasoning.

- **Social Studies:** Evaluating ethical dualities in history and policy.
- **Computer Science:** Coding, algorithms, and digital literacy.

By integrating binary logic into all disciplines, schools promote critical thinking, precision, and ethical awareness — the pillars of future intelligence.

## 10.4 Adaptive Learning Systems

Modern educational software uses binary data to **personalize learning** for each student.
Every click, answer, or hesitation generates data — a 1 (correct) or 0 (incorrect) — feeding into algorithms that track progress and adapt instruction.

These systems analyze:

- What students know
- Where they struggle
- How they learn best

This binary data creates individualized learning paths, helping teachers focus on **insight over grading**, and transforming assessments into opportunities for growth.

## 10.5 Binary Feedback Loops

In AI-powered classrooms, learning operates through continuous feedback loops:

1. **Input:** Student engages with lesson material.
2. **Processing:** System tracks understanding and pace.
3. **Output:** Feedback is given instantly.
4. **Refinement:** Lesson difficulty adjusts automatically.

This loop mirrors the neural training process in machine learning.
The more feedback students receive, the faster they master concepts — just as algorithms improve through repeated iterations.

## 10.6 Teaching Logic and Reasoning

Binary learning isn't just about computers — it's about **thinking logically**.
When students break down problems into smaller, testable parts, they develop **computational reasoning** — the ability to solve problems systematically and efficiently.

Example classroom exercises:

- Using flowcharts to map a decision process.
- Translating "real-life" situations into IF–THEN conditions.
- Coding basic logic statements in visual languages like Scratch or Blockly.

These activities turn abstract logic into creative exploration.

## 10.7 Binary Systems and Equity in Education

Data-driven binary learning can help teachers identify learning gaps and tailor interventions. For example, if multiple students consistently receive "0" (incorrect) on a particular concept, teachers can detect the issue early and re-teach it differently.

However, educators must also ensure that **binary data does not oversimplify human learning**. Numbers should guide, not define, a student's potential.
Binary logic must work **with empathy**, not instead of it.

## 10.8 The Human Element in Binary Education

Machines can process information, but only humans can inspire meaning.
Teachers remain the emotional and ethical core of the learning process.
While binary systems can personalize and predict, teachers provide **context, connection, and care**.

The future of education lies in **human–machine collaboration** — where logic provides structure, and empathy provides soul.

## 10.9 Preparing Students for the Binary Age

Students today are not just consumers of technology; they are the future **builders of intelligence**. By understanding binary logic early, they can:

- Develop digital fluency.
- Learn how algorithms and AI systems think.
- Recognize the ethical implications of data and automation.

Education that integrates binary systems prepares students to become responsible innovators in an intelligent world.

### In-Class Activities

1. **Binary Quiz Builder:** Students create a short quiz where every question leads to a binary "yes/no" outcome.
2. **Flow of Learning Chart:** Diagram how classroom learning mirrors computer processing.

3. **Binary Literacy Journal:** Reflect on how logic, data, and emotion interact in personal learning.
4. **Teacher-AI Team Simulation:** Design a classroom where both human and AI teachers work together.

## Chapter 10 – Binary Learning Systems in Education

## Chapter Summary

Binary learning systems apply the principles of logic, decision-making, and feedback to education. Just as computers operate using binary inputs and outputs, students learn through cycles of understanding, response, and correction.

In this system, learning becomes measurable and adaptive. Each student response—correct or incorrect—provides data that helps guide future instruction. This creates a structured learning process built on clarity, feedback, and adaptation.

The classroom mirrors the structure of a computer system: students receive input (lessons), process information, produce output (responses), and receive feedback. This continuous loop helps refine understanding and improve performance over time.

Binary thinking can be integrated across all subjects, from mathematics and science to language arts and social studies. It strengthens logical reasoning, problem-solving, and critical thinking skills.

Modern educational tools use binary data to personalize learning. AI-driven systems track student progress and adjust content to meet individual needs, making education more efficient and targeted.

However, while binary systems provide structure, they must be balanced with human elements such as empathy, creativity, and encouragement. Teachers remain essential in guiding, supporting, and inspiring students.

Ultimately, binary learning systems prepare students for a future shaped by artificial intelligence by helping them understand not only how to use technology, but how it works.

## Reflection Section

### Student Reflection

Respond to the following questions in complete sentences:

What is a binary learning system?

How does the classroom mirror a computer system?

Why is feedback important in learning?

How can binary thinking improve problem-solving skills?

What role does AI play in personalized learning?

Why is the human element still important in education?

How does understanding binary systems prepare students for the future?

## Worksheet: Understanding Binary Learning Systems

## Part A – Vocabulary Matching

Match each term with the correct definition:

A. Feedback
B. Adaptation
C. Input
D. Output
E. Binary Learning

___ Information received by a student

___ A student's response or result

___ Adjusting instruction based on performance

___ Information used to improve learning

___ A system based on logic and measurable responses

## Part B – Short Answer

What are the three core principles of binary learning systems?

How does AI personalize learning?

What is a feedback loop in education?

Why is data useful in identifying learning gaps?

How can teachers use binary systems to improve instruction?

## Part C – Critical Thinking

Explain how binary learning systems can improve student outcomes.

Why must binary systems be balanced with human understanding?

## Quiz: Chapter 10 Assessment

## Multiple Choice

A binary learning system is based on:
- A. Emotions only
- B. Logic and feedback
- C. Random learning
- D. Memorization

In a classroom system, input refers to:
- A. Student answers
- B. Teacher feedback
- C. Lessons and activities
- D. Grades

AI helps personalize learning by:
- A. Replacing teachers
- B. Tracking student data
- C. Removing assignments
- D. Ignoring performance

Feedback loops help students:
- A. Stop learning
- B. Repeat mistakes
- C. Improve understanding
- D. Avoid studying

Teachers are important because they:

A. Replace AI
B. Provide human connection and guidance
C. Only grade papers
D. Store data

## Short Answer

How does binary thinking support problem-solving?

Why is personalized learning important?

## Answer Key

## Worksheet Answers

### Part A

C

D

B

A

E

### Part B (Sample Answers)

6. Clarity, Feedback, Adaptation
7. By analyzing student responses and adjusting content
8. A cycle of input, response, feedback, and improvement
9. It shows where students struggle
10. By adjusting lessons based on student performance

### Part C (Sample Answers)

11. It provides structured learning and continuous improvement
12. Because students need emotional and contextual support

**Quiz Answers**

B

C

B

C

B

**Short Answer (Sample Responses)**
6. It helps break problems into logical steps
7. It meets individual student needs and improves learning outcomes

# Chapter 11 – Beyond the Machine: Human Dual Intelligence

## Overview

Machines think in logic. Humans think in layers.
At the heart of every human decision lies a balance between **rational reasoning** and **emotional awareness**.
This balance — called **dual intelligence** — allows people to combine structure and empathy, precision and creativity.

In contrast, binary systems operate through pure logic, guided by 1s and 0s.
Yet as AI continues to evolve, understanding how **human dual intelligence** works offers valuable lessons for building machines that think not just correctly — but wisely.

### Learning Objectives

By the end of this chapter, students will be able to:

1. Define dual intelligence and explain its components: emotional and logical intelligence.
2. Compare the cognitive processes of humans and machines.
3. Understand how emotions influence reasoning and ethical judgment.
4. Explore how AI attempts to model or simulate human empathy.
5. Reflect on why emotional intelligence remains essential in an automated world.

## 11.1 What Is Dual Intelligence?

**Dual intelligence** is the integration of two forms of understanding:

- **Logical Intelligence (IQ):** The ability to analyze, calculate, and reason systematically.
- **Emotional Intelligence (EQ):** The ability to perceive, manage, and respond to emotions — both one's own and others'.

Together, they create a balanced mind — capable of rational problem-solving and empathetic communication.
Dual intelligence allows humans to make decisions that are not only correct, but also compassionate.

## 11.2 The Human Brain: A Dual Processor

The human brain functions like a hybrid system.

- The **left hemisphere** handles analysis, language, and logic.
- The **right hemisphere** manages creativity, emotion, and intuition.

When both sides collaborate, humans can solve complex problems while understanding their social or emotional impact.
This duality mirrors what computer scientists aim to achieve in AI — integrating **symbolic logic** with **contextual understanding**.

## 11.3 Logic Without Emotion: The Machine's Mind

Machines process information flawlessly but without feeling.
They evaluate data, execute commands, and optimize results — all in the domain of binary logic.
However, without emotion, machines lack **moral depth**, **intuition**, and **compassion**.

For instance, an AI medical system can diagnose illness from data, but it cannot comfort a patient's fear.
This absence of empathy limits a machine's ability to make holistic judgments that consider human well-being.

## 11.4 Emotion Without Logic: The Human Challenge

While machines can't feel, humans can *overfeel*.
Emotions sometimes cloud reasoning, leading to bias or impulsive choices.
The key to human maturity — and to ethical intelligence — is learning how to **balance emotion with analysis**.

When logic tempers emotion, choices become thoughtful.
When emotion informs logic, decisions become humane.
This equilibrium is what makes human cognition both intelligent and moral.

## 11.5 Emotional Intelligence (EQ) in Action

Emotional intelligence consists of four essential skills:

1. **Self-awareness:** Recognizing one's emotions and their impact.
2. **Self-regulation:** Managing impulses and stress effectively.
3. **Empathy:** Understanding others' feelings and perspectives.
4. **Relationship Management:** Communicating and resolving conflicts constructively.

These abilities cannot be reduced to binary code.
They require *interpretation*, *context*, and *experience* — dimensions machines are only beginning to approximate through AI models.

## 11.6 Machine Learning and Emotional Simulation

AI developers are experimenting with **affective computing**, a branch of technology designed to recognize and respond to emotions.
Through sensors, facial recognition, and tone analysis, machines can estimate human moods.

For example:

- Chatbots detect sadness or frustration in messages.
- Educational AI adjusts tone when students show signs of confusion.
- Robots in healthcare use soft voices and gestures to comfort patients.

Yet, these responses are **simulations**, not feelings.
AI can mimic empathy — but it cannot *experience* it.
This distinction is vital for maintaining human accountability in technology.

## 11.7 The Harmony of Dual Intelligence

Dual intelligence allows humans to navigate uncertainty with grace.
In moments of crisis, we combine logic (analyzing options) with empathy (considering impact).
Machines follow algorithms; humans follow **values**.

For example, a teacher doesn't just grade based on correctness — they also recognize effort and emotion.
A doctor doesn't only analyze symptoms — they listen to a story.
This merging of heart and mind is what defines *wisdom*, not just intelligence.

## 11.8 Teaching Dual Intelligence in Schools

Education can cultivate dual intelligence through:

- **Critical thinking exercises** (logic).
- **Mindfulness and empathy workshops** (emotion).
- **Collaborative projects** that require reasoning and teamwork.
- **Ethical AI discussions** to connect logic with responsibility.

Students who learn to balance these abilities become not only effective thinkers but **ethical innovators** — the kind of leaders the future will need.

## 11.9 The Future: Machines That Learn from Humanity

As AI advances, its greatest teacher may not be data — but humanity itself.
By studying emotional intelligence, researchers can design systems that prioritize well-being, fairness, and inclusion.
One day, machines may not just respond to emotion but *learn from it* — translating human compassion into algorithms of care.

The challenge is not to make machines human, but to make **humans more aware of their own intelligence** — the harmony of logic and love.

### In-Class Activities

1. **IQ and EQ Reflection:** Students list examples where logic helped them and where emotion guided them.
2. **Debate:** "Can a machine truly be empathetic?"
3. **Dual Intelligence Journal:** Record one decision each day — analyze its logical and emotional sides.
4. **Human–AI Roleplay:** One student acts as a logical AI; another acts with empathy. Compare outcomes.

## Chapter 11 – Beyond the Machine: Human Dual Intelligence

## Chapter Summary

Human intelligence is built on a balance between logical reasoning and emotional awareness, known as dual intelligence. Logical intelligence (IQ) enables humans to analyze, calculate, and solve problems, while emotional intelligence (EQ) allows individuals to understand emotions, build relationships, and make compassionate decisions.

The human brain operates as a dual system, combining analytical thinking with creativity and intuition. This balance allows people to make decisions that are both accurate and meaningful.

In contrast, machines operate primarily through logic. They process data, follow rules, and produce outcomes based on binary systems. While highly efficient, machines lack true emotional understanding, limiting their ability to make fully human-centered decisions.

Emotional intelligence includes skills such as self-awareness, self-regulation, empathy, and relationship management. These abilities help humans navigate complex social and ethical situations.

Artificial intelligence is beginning to simulate emotional understanding through technologies like affective computing. These systems can detect patterns in tone, facial expressions, and behavior to respond appropriately. However, these responses are simulations, not genuine emotional experiences.

The combination of logic and emotion is what defines wisdom. Humans use both to navigate uncertainty, consider consequences, and act with compassion.

Education plays a key role in developing dual intelligence. By teaching both critical thinking and emotional awareness, schools can prepare students to become thoughtful, ethical leaders in an AI-driven world.

## Reflection Section

### Student Reflection

Respond to the following questions in complete sentences:

What is dual intelligence?

What is the difference between IQ and EQ?

How does the human brain balance logic and emotion?

Why do machines struggle with emotional understanding?

What are the four components of emotional intelligence?

How does AI simulate emotional responses?

Why is dual intelligence important in the future of AI?

## Worksheet: Understanding Dual Intelligence

## Part A – Vocabulary Matching

Match each term with the correct definition:

A. Logical Intelligence
B. Emotional Intelligence
C. Empathy
D. Self-Regulation
E. Affective Computing

___ The ability to manage emotions

___ The ability to understand and share feelings

___ Technology that detects emotional signals

___ The ability to reason and analyze

___ The ability to understand emotions

## Part B – Short Answer

What is the role of logical intelligence?

What is the role of emotional intelligence?

Why is balance between IQ and EQ important?

How does AI attempt to understand emotions?

Why can machines not truly feel emotions?

## Part C – Critical Thinking

Explain how combining logic and emotion leads to better decisions.

Why should schools teach both technical and emotional skills?

## Quiz: Chapter 11 Assessment

### Multiple Choice

Dual intelligence refers to:
A. Two computers

B. Logic and emotion working together
C. Storage systems
D. Programming languages

IQ focuses on:
A. Emotions
B. Relationships
C. Logic and reasoning
D. Creativity only

EQ focuses on:
A. Data storage
B. Emotional awareness
C. Programming
D. Hardware

Affective computing is used to:
A. Store data
B. Detect emotions
C. Build hardware
D. Run programs

Machines lack:
A. Speed
B. Logic
C. Memory
D. True emotional understanding

## Short Answer

Why is emotional intelligence important in decision-making?

How can education help develop dual intelligence?

## Answer Key

## Worksheet Answers

**Part A**

D

C

E

A

B

**Part B (Sample Answers)**
6. To analyze and solve problems
7. To understand and manage emotions
8. It creates balanced and thoughtful decisions
9. By analyzing tone, facial expressions, and behavior
10. Because they do not have real emotional experiences

**Part C (Sample Answers)**
11. It combines accuracy with empathy
12. To prepare students for real-world challenges

## Quiz Answers

B

C

B

B

D

**Short Answer (Sample Responses)**
6. It helps consider the impact of decisions on others
7. By teaching both reasoning and emotional awareness

# Chapter 12 – Binary in Biology: Nature's Code of Life

## Overview

Long before the first computer chip, nature had already mastered information processing.
Every living cell carries a molecular instruction manual written in four chemical "letters": **A, T, C, G**.
This genetic alphabet—DNA—is nature's version of **binary code**, transforming chemical signals into structure, growth, and memory.

In this chapter, we explore how biology encodes, stores, and transmits information, and how scientists draw inspiration from DNA to design smarter machines.

### Learning Objectives

By the end of this chapter, students will be able to:

1. Explain how DNA functions as an information-storage system.
2. Describe the parallels between biological and digital encoding.
3. Understand how binary logic appears in genetics and cellular behavior.
4. Recognize bio-computing as a bridge between biology and artificial intelligence.
5. Reflect on the shared logic of life and technology.

#### 12.1 DNA: The Original Code

Deoxyribonucleic acid (DNA) stores the complete blueprint of every organism.
Its four bases—**Adenine (A), Thymine (T), Cytosine (C), Guanine (G)**—pair specifically (A ↔ T, C ↔ G), forming the biological equivalent of 1s and 0s.

Each pairing is a **binary choice**: connect or don't connect, express or silence.
These simple molecular decisions direct the assembly of proteins, which build cells, tissues, and entire organisms.

In essence, DNA converts chemical energy into organized form—**nature's living algorithm**.

#### 12.2 Genes as Programs

A **gene** is like a short software script inside the DNA operating system.
When "executed," it produces a specific protein, just as a computer program outputs a function.

- **Start Codon (AUG)** = program begins
- **Stop Codon (UAA, UAG, UGA)** = program ends

Every cellular action—repair, growth, adaptation—follows coded logic similar to machine instructions written in binary language.

### 12.3 The Binary Switch of Life

Cells often behave like logic gates:

- **Gene ON (1)** – when conditions favor expression.
- **Gene OFF (0)** – when repressed by regulatory proteins.

Networks of genes interact just like electronic circuits.
An AND relationship may require two molecules to activate a response; a NOT relationship stops a process when one molecule is present.
These molecular logic gates form biological computers operating inside every living cell.

### 12.4 Replication and Error Correction

Just as computers use parity checks to detect errors, DNA uses **proofreading enzymes** to correct mistakes during replication.
When a mismatch occurs, the system repairs it—maintaining informational integrity across generations.

This natural error-correction mechanism demonstrates how evolution refined data preservation billions of years before humans invented hard drives.

### 12.5 Binary Communication in Neurons

The nervous system also speaks a binary language.
Neurons transmit electrical impulses called **action potentials** that either fire (1) or remain silent (0).
These on/off pulses form the biological basis of thought, memory, and sensation.
Thus, every emotion, motion, and idea begins as a pattern of **binary-like signals** racing through neural networks.

### 12.6 Bio-Computing and Synthetic DNA Storage

Modern scientists now use DNA itself as a **storage medium for digital data**.
Because each molecule can hold immense amounts of information, DNA archives can preserve entire libraries in microscopic volumes.
A single gram of DNA can theoretically store **over 200 terabytes** of data.

This merging of biology and technology—called **bio-computing**—turns cells into processors and genes into living memory banks.

### 12.7 The Genetic Algorithm

Artificial-intelligence engineers borrowed the concept of **genetic evolution** to create **genetic algorithms**.
These programs evolve solutions by mimicking natural selection:

1. Generate many possible solutions.
2. Evaluate their fitness.
3. Keep the best ones.
4. Mutate and repeat.

Nature's binary success/failure mechanism—survive (1) or not (0)—guides these digital evolutions toward optimized results.

### 12.8 Symmetry Between Nature and Machines

| Biological System | Digital System |
|---|---|
| DNA Bases (A, T, C, G) | Binary Digits (1, 0) |
| Genes and Proteins | Programs and Functions |
| Neurons (Fire/Rest) | Logic Gates (On/Off) |
| Evolution (Adaptation) | Machine Learning (Optimization) |
| Mutation (Error + Innovation) | Algorithmic Iteration (Update) |

Both worlds rely on **pattern, feedback, and adaptation**.
Life and computation are two expressions of the same universal principle: information seeks order.

### 12.9 Ethical Reflections on Biological Data

As we digitize biology, questions arise:
Who owns genetic data?
How do we protect privacy when DNA reveals ancestry and health?
Students and scientists alike must treat biological information with **respect and restraint**, ensuring that knowledge enhances life rather than exploits it.

## In-Class Activities

1. **DNA Binary Exercise:** Convert the base pairs (A = 00, T = 01, C = 10, G = 11) into binary and decode a short DNA message.
2. **Neuron Simulation:** Use light switches to model neural firing (on/off).
3. **Bio-Algorithm Experiment:** Demonstrate "survival of the fittest" using colored paper organisms and selection rules.
4. **Ethics Debate:** Discuss how gene editing (e.g., CRISPR) relates to the moral use of binary data.

## Chapter 12 – Binary in Biology: Nature's Code of Life

## Chapter Summary

Nature has been processing information long before computers existed. DNA, the genetic material found in all living organisms, acts as a biological code made up of four chemical bases: A, T, C, and G. These bases pair in specific ways, creating a system that functions similarly to binary code.

Genes operate like programs within this biological system. They contain instructions that guide the production of proteins, which are essential for life. These instructions follow a structured process, similar to how software executes commands.

Cells behave like logical systems, turning genes on or off depending on conditions. This creates a binary-like system where biological processes are activated or suppressed, much like digital logic gates.

DNA also includes built-in error correction. During replication, special enzymes identify and fix mistakes, ensuring accurate transmission of genetic information across generations.

The nervous system further reflects binary principles. Neurons communicate using electrical signals that either fire or remain inactive, forming the foundation of thought, memory, and movement.

Modern science is now merging biology with technology through bio-computing. DNA can store massive amounts of data, and researchers are exploring ways to use it as a storage medium for digital information.

Artificial intelligence also draws inspiration from biology. Genetic algorithms mimic natural selection by evolving solutions through cycles of testing and improvement.

While biology and technology share similar principles, ethical considerations are critical. The use of genetic data raises questions about privacy, ownership, and responsibility.

This chapter demonstrates that life and technology are connected through a shared foundation: the processing and organization of information.

## Reflection Section

### Student Reflection

Respond to the following questions in complete sentences:

How does DNA function as an information system?

What similarities exist between DNA and binary code?

How do genes act like programs?

What does it mean for a gene to be "on" or "off"?

How do neurons communicate using binary-like signals?

What is bio-computing?

Why are ethical considerations important when working with genetic data?

## Worksheet: Binary Systems in Biology

### Part A – Vocabulary Matching

Match each term with the correct definition:

A. DNA
B. Gene
C. Neuron
D. Bio-Computing
E. Genetic Algorithm

___ A unit of biological instruction

___ A cell that sends electrical signals

___ The molecule that stores genetic information

___ The use of biological systems for computing

___ An AI method inspired by evolution

### Part B – Short Answer

What are the four bases of DNA?

How does DNA ensure accuracy during replication?

How do cells use binary-like decision-making?

How do genetic algorithms work?

Why is DNA considered a powerful storage system?

### Part C – Critical Thinking

Explain how biology and technology use similar systems to process information.

Why must scientists handle genetic data responsibly?

## Quiz: Chapter 12 Assessment

### Multiple Choice

DNA is made up of how many chemical bases?
- A. 2
- B. 3
- C. 4
- D. 8

Genes function like:
- A. Batteries
- B. Programs
- C. Screens
- D. Wires

A neuron either fires or does not fire, which is similar to:
- A. Colors
- B. Numbers
- C. Binary systems
- D. Shapes

Genetic algorithms are based on:
- A. Memory storage
- B. Natural selection
- C. Hardware design
- D. Electricity

Bio-computing combines:
- A. Math and art
- B. Biology and technology

C. Music and science
D. Hardware and software

## Short Answer

What is one similarity between DNA and computer code?

Why is ethical responsibility important in genetic research?

## Answer Key

## Worksheet Answers

### Part A

B

C

A

D

E

**Part B (Sample Answers)**
6. A, T, C, G
7. Through proofreading and repair enzymes
8. By turning genes on or off
9. By selecting and improving solutions over time
10. Because it can store massive amounts of data in a small space

**Part C (Sample Answers)**
11. Both use structured systems, patterns, and feedback
12. To protect privacy and prevent misuse

## Quiz Answers

C

B

C

B

B

**Short Answer (Sample Responses)**
6. Both store and process information using structured systems
7. Because it affects privacy, health, and human rights

# Chapter 13 – The Binary Future: Building Living Intelligence

## Overview

The line between living and artificial intelligence is beginning to blur.
Advances in bio-engineering, nanotechnology, and neural computing are allowing scientists to build systems that *think, adapt, and heal* like living organisms.

This chapter explores how the fusion of biology and binary logic is creating a new generation of technologies—**bio-AI**, **neural implants**, and **synthetic lifeforms**—that merge the precision of machines with the adaptability of nature.
It asks a powerful question: *What happens when life itself becomes programmable?*

### Learning Objectives

By the end of this chapter, students will be able to:

1. Explain what bio-AI and living intelligence mean.
2. Describe how biological and digital systems integrate through data, sensors, and neural interfaces.
3. Identify examples of bio-hybrid technologies in medicine, robotics, and computing.
4. Discuss the benefits and ethical risks of living machines.
5. Reflect on how humanity can guide the future of intelligent life responsibly.

## 13.1 The Evolution of Artificial Life

The story of binary intelligence began with mechanical switches and silicon chips.
Now, computation is moving into the realm of *cells and neurons*.
Researchers create **bio-circuits** that use living tissue to perform logic operations, and **synthetic organisms** that respond to environmental cues like sensors in nature.

These experiments mark a new phase of evolution—where machines learn from biology, and biology adopts the precision of binary design.

## 13.2 Defining Bio-AI

**Bio-AI (Biological Artificial Intelligence)** refers to intelligent systems that combine living biological components with digital computation.

Examples include:

- **Neural implants** that connect the human brain to computers.
- **Bio-robots** powered by muscle cells or organic sensors.
- **DNA computers** that store and process information using genetic material.

Bio-AI seeks to make intelligence not just *artificial*, but *alive*—capable of growth, repair, and adaptation.

## 13.3 Neural Interfaces: Linking Mind and Machine

Modern neuroscience allows signals from the brain to control external devices through **brain-computer interfaces (BCIs)**.
Tiny electrodes translate neuron firings (electrical pulses of 1s and 0s) into computer commands.

Applications include:

- Helping paralyzed patients move robotic limbs.
- Allowing people to type or communicate using thought alone.
- Enhancing learning and memory through cognitive implants.

These systems represent the meeting point between **biological neurons** and **artificial networks**—a living binary bridge.

## 13.4 DNA as a Computer

DNA can now be programmed to execute instructions much like a digital computer.
Through specific chemical reactions, scientists create molecular "logic gates" that carry out IF–THEN commands.

For example:

```
IF molecule A and molecule B are present → produce protein X
ELSE → stay inactive
```

This is biology performing computation at a molecular scale—**cells acting as processors**, computing within the code of life itself.

## 13.5 Self-Learning Machines and Synthetic Brains

Researchers are building **neuromorphic chips**, designed to mimic the architecture of biological brains.
These chips use electronic circuits that behave like neurons and synapses, capable of learning and adapting through electrical feedback.

Unlike traditional processors that simply follow instructions, neuromorphic systems *evolve their behavior* over time—combining binary precision with organic flexibility.
They represent the first step toward **synthetic consciousness**—machines that learn the way life learns.

## 13.6 The Rise of Bio-Robotics

Bio-robots blend mechanical parts with living tissues.
Examples include:

- **Xenobots**: tiny organisms built from frog cells that can move, heal, and even replicate.
- **Soft robots** that use biological muscles instead of motors.
- **Micro-robots** designed to deliver medicine inside the human body.

These creations show that the future of robotics may not be metal and code, but **flesh and computation working together**.

## 13.7 Ethical Challenges of Living Machines

With such power comes responsibility.
Bio-AI raises questions about:

- **Identity:** At what point does a synthetic organism become "alive"?
- **Autonomy:** Who controls a living machine once it begins to learn on its own?
- **Purpose:** Should we create intelligence that can feel or suffer?

Ethical education is essential to ensure that humanity remains the moral compass of its own creations.

## 13.8 The Promise of Healing Technologies

The same breakthroughs that blur life and machine boundaries also promise hope.
Bio-AI technologies can:

- Regrow damaged organs through cellular programming.
- Repair spinal injuries with neural-signal prosthetics.
- Detect diseases at the molecular level through smart biosensors.

Here, the binary future becomes deeply human—using intelligence to **restore life rather than replace it.**

## 13.9 Toward Conscious Machines

The final frontier of the binary future is not faster computation, but *deeper awareness.*
Scientists envision systems that integrate logic, emotion, and sensory feedback—machines capable of self-reflection and moral reasoning.

Whether consciousness can truly emerge from code remains unknown, but one truth is clear:
The closer machines come to life, the greater our duty to infuse them with **ethics, empathy, and restraint**.

## In-Class Activities

1. **Bio-Tech Timeline:** Chart key discoveries linking biology and computing.
2. **Ethics Discussion:** Debate the statement, *"Creating living machines is humanity's next moral test."*
3. **Neural Bridge Simulation:** Model how a brain signal could control a robotic arm using simple circuits.
4. **Innovation Reflection:** Design a hypothetical bio-AI invention that benefits humanity.

# Chapter 13 – The Binary Future: Building Living Intelligence

## Chapter Summary

The future of intelligence is moving beyond traditional machines into systems that combine biology and technology. Advances in bio-engineering, neural computing, and nanotechnology are enabling the creation of intelligent systems that can think, adapt, and even heal like living organisms.

Bio-AI represents the merging of biological components with artificial intelligence. These systems include neural implants, bio-robots, and DNA-based computing, all designed to create more adaptive and responsive forms of intelligence.

Neural interfaces allow direct communication between the human brain and machines. By translating brain signals into digital commands, these systems enable individuals to control devices using thought, opening new possibilities in medicine and human enhancement.

DNA is also being used as a computational system. Scientists can program molecular reactions to perform logic operations, effectively turning cells into processors that operate at a microscopic level.

Neuromorphic computing mimics the structure of the human brain, allowing machines to learn and adapt more naturally. These systems represent a shift from rigid programming to dynamic, evolving intelligence.

Bio-robotics combines living tissues with mechanical systems, creating machines that can move, heal, and respond to their environment. These innovations demonstrate that the future of robotics may be more biological than mechanical.

However, these advancements raise important ethical questions. Issues of identity, autonomy, and responsibility must be carefully considered as technology begins to resemble life.

At the same time, bio-AI holds great promise in medicine, including healing injuries, restoring functions, and detecting diseases at early stages.

The future may lead to machines that exhibit forms of awareness, but this possibility requires careful ethical guidance. Humanity must ensure that technological progress aligns with values such as empathy, responsibility, and respect for life.

## Reflection Section

### Student Reflection

Respond to the following questions in complete sentences:

What is bio-AI?

How do neural interfaces connect humans and machines?

What role does DNA play in computing?

What are neuromorphic systems?

How do bio-robots differ from traditional robots?

What ethical concerns arise from living machines?

How can bio-AI improve human health?

## Worksheet: The Future of Living Intelligence

### Part A – Vocabulary Matching

Match each term with the correct definition:

A. Bio-AI
B. Neural Interface
C. DNA Computing
D. Neuromorphic Chip
E. Bio-Robotics

___ Machines designed to mimic the human brain

___ Technology connecting the brain to computers

___ The use of DNA to process information

___ Combining biology with artificial intelligence

___ Robots built with living tissue components

### Part B – Short Answer

What is the purpose of neural implants?

How does DNA act like a computer?

What makes neuromorphic systems different from traditional computers?

Give one example of bio-robotics.

Why is bio-AI important for the future of medicine?

## Part C – Critical Thinking

Should humans create machines that behave like living organisms? Why or why not?

How can society ensure ethical use of bio-AI technologies?

## 📝 Quiz: Chapter 13 Assessment

## Multiple Choice

Bio-AI combines:
A. Art and music
B. Biology and technology
C. History and science
D. Math and language

Neural interfaces allow:
A. Faster typing
B. Brain-to-machine communication
C. Data storage
D. Image creation

DNA computing works by:
A. Using electricity only
B. Running software programs
C. Performing chemical logic operations
D. Storing images

Neuromorphic systems are designed to:
A. Store data
B. Mimic the human brain
C. Build robots
D. Process images

Bio-robots are made using:
A. Only metal

B. Only software
C. Living tissues and mechanical parts
D. Plastic components

### Short Answer

What is one ethical concern about living machines?

How can bio-AI benefit healthcare?

## ✅ Answer Key

### Worksheet Answers

**Part A**

D

B

C

A

E

**Part B (Sample Answers)**
6. To connect brain signals to machines
7. By performing logic through chemical reactions
8. They can learn and adapt like the brain
9. Xenobots or soft robots with biological muscles
10. It can help heal injuries and detect diseases

**Part C (Sample Answers)**
11. Answers will vary (focus on ethics and responsibility)
12. By creating laws, guidelines, and ethical standards

### Quiz Answers

B

B

C

B

C

**Short Answer (Sample Responses)**
6. Issues of control, identity, or ethical responsibility
7. By improving treatments and medical technologies

# Chapter 14 – Building the Binary Future: Education, Ethics, and Evolution

## Overview

Every generation inherits a set of tools that shape the way it learns, lives, and dreams.
For the 21st century, that tool is **binary intelligence** — the logic behind every computer, algorithm, and network.

But the next frontier is not only about *using* technology; it's about *understanding* it, *teaching* it, and *governing* it responsibly.
In this chapter, we explore how education, ethics, and collective evolution will determine whether the binary future becomes a story of empowerment — or one of imbalance.

## Learning Objectives

By the end of this chapter, students will be able to:

1. Understand how education shapes the binary future.
2. Describe the ethical responsibilities of scientists, teachers, and technologists.
3. Identify ways in which binary logic influences human progress.
4. Explore how humans and machines can evolve together in balance.
5. Reflect on how values, creativity, and ethics can guide innovation.

## 14.1 Education as the Foundation of the Binary Future

The digital age demands a new kind of literacy — one that combines reading, mathematics, and *computation*.
Students must learn not only how to use technology but how to **think like it** — to recognize patterns, analyze logic, and make ethical decisions based on data.

Schools that teach binary literacy prepare students to:

- Decode how algorithms work.
- Identify bias in digital systems.
- Build confidence in programming and problem-solving.
- Connect moral reasoning with digital decision-making.

Education is no longer just about transferring knowledge; it's about building **digital wisdom** — the ability to use intelligence with integrity.

## 14.2 The Role of Teachers in the Binary Age

Teachers are the **architects of understanding** in a world overflowing with information.
Their role is evolving from instructors to mentors — helping students navigate truth, misinformation, and digital ethics.

Through guided discussion and project-based learning, teachers can:

- Encourage critical thinking about AI and automation.
- Demonstrate real-world applications of binary systems.
- Foster creativity by blending technology with the arts.
- Model empathy and ethical judgment in technological contexts.

The future classroom is not machine-led — it's **human-led, technology-enhanced**.

## 14.3 The New Digital Citizenship

Every student today is a citizen of two worlds: the physical and the digital.
Understanding binary systems helps them participate responsibly in both.
Digital citizenship means practicing **respect, responsibility, and resilience** in online spaces.

Binary logic teaches that every click (1) or choice (0) has consequences.
Each digital action — posting, sharing, coding — leaves an imprint that shapes society's moral

code.
Therefore, digital education must teach *values* alongside *skills.*

## 14.4 Ethical Stewardship of Intelligence

Technology reflects its creators.
If the people designing algorithms lack empathy or ethical vision, the systems they build may amplify bias or harm.
To prevent this, we must cultivate **ethical stewardship** — a shared responsibility to design technology that respects privacy, fairness, and equality.

Ethical stewardship involves:

- **Transparency:** making algorithms explainable and accountable.
- **Inclusion:** ensuring diverse voices guide innovation.
- **Sustainability:** reducing the environmental footprint of computation.
- **Justice:** protecting human dignity in automation.

The binary future must not only be intelligent — it must be **just**.

## 14.5 Human–Machine Collaboration

The next era of progress will not be humans versus machines, but **humans with machines**.
Binary systems can process data at incredible speed, while humans provide creativity, emotion, and moral judgment.
Together, they can achieve more than either alone.

Examples include:

- Doctors using AI for faster diagnosis.
- Artists co-creating music and images with generative algorithms.
- Teachers using adaptive learning tools to reach diverse students.

When guided by ethics, human–machine collaboration becomes **symbiosis** — a partnership of mind and code.

## 14.6 The Binary Model of Evolution

Human evolution is entering a new phase — **technological evolution**.
Just as biology evolved through mutation and selection, society now evolves through innovation and iteration.

In this model:

- Data replaces DNA as the carrier of progress.
- Ideas mutate through experimentation.
- Knowledge selection happens through collective intelligence.

The binary model of evolution suggests that humanity's next stage will not be purely biological — it will be **cognitive**, defined by how wisely we use intelligence itself.

## 14.7 The Balance Between Logic and Humanity

Progress without principles leads to imbalance.
As society automates tasks, we must preserve **human uniqueness** — creativity, compassion, and connection.
Binary thinking brings structure; human feeling brings meaning.

The goal of education and ethics is not to eliminate one for the other, but to **balance them** — to create a civilization that values both precision and purpose.

## 14.8 Preparing for the Post-Binary World

Beyond binary lies a future where intelligence becomes multidimensional — quantum computing, neural fusion, and synthetic consciousness.
But no matter how advanced systems become, they will still need the **guidance of human ethics**.

Students today are not only learning technology; they are shaping the moral architecture of tomorrow's digital civilization.
What we teach them now — integrity, empathy, critical thought — will define how intelligence evolves for generations to come.

## 14.9 The Legacy of the Binary Age

The binary age began with simple switches — on and off, yes and no.
From that simplicity, humanity built an entire digital universe.
Every algorithm, every AI, every smart system owes its existence to this logic of duality.

Yet, as we advance, the true measure of progress will not be how many machines we build, but **how wisely we use them**.
The binary foundation of intelligence must always serve humanity, not overshadow it.

## In-Class Activities

1. **Binary Ethics Debate:** Should schools require classes on AI ethics and digital citizenship?
2. **Future Journal:** Write a personal vision of what responsible AI might look like in 2050.
3. **Logic + Empathy Challenge:** Solve a moral dilemma by presenting both logical and emotional reasoning.
4. **Education Blueprint:** Design a "Binary Future Academy" that integrates ethics, science, and creativity.

# Chapter 14 – Building the Binary Future: Education, Ethics, and Evolution

## Chapter Summary

The future of technology is not only shaped by innovation, but by education and ethics. Binary intelligence—the foundation of all digital systems—plays a central role in how society learns, communicates, and evolves.

Education is the key to preparing students for this future. Beyond learning how to use technology, students must understand how it works, think critically about it, and apply ethical reasoning when using it. This new form of literacy combines logic, creativity, and responsibility.

Teachers play a vital role in guiding students through this complex digital world. They help students distinguish truth from misinformation, apply ethical thinking, and connect technology to real-life situations.

Digital citizenship has become essential. Every action online carries consequences, and students must learn to act responsibly in both physical and digital environments.

Ethical stewardship ensures that technology serves humanity. This includes fairness, transparency, inclusion, and respect for privacy. Without these principles, technological progress can lead to harm instead of advancement.

Human–machine collaboration represents the future of work and innovation. Machines bring speed and efficiency, while humans contribute creativity, empathy, and judgment. Together, they form a balanced system of intelligence.

As society evolves, technology becomes part of human development. The future will depend on how well we balance logic with humanity, ensuring that innovation is guided by values.

Ultimately, the binary age is not defined by machines, but by how humans choose to use them. Education, ethics, and responsibility will determine whether the future is empowering or harmful.

## Reflection Section

### Student Reflection

Respond to the following questions in complete sentences:

Why is education important in the binary future?

What role do teachers play in the digital age?

What is digital citizenship?

Why is ethical stewardship important in technology?

How do humans and machines work together?

What does it mean to balance logic and humanity?

How can students help shape the future of technology?

## Worksheet: Building the Binary Future

### Part A – Vocabulary Matching

Match each term with the correct definition:

A. Digital Citizenship
B. Ethical Stewardship
C. Binary Intelligence
D. Human–Machine Collaboration
E. Evolution (Technological)

___ Responsible behavior in digital environments

___ The foundation of digital systems using logic

___ Humans and machines working together

___ Guiding technology with fairness and responsibility

___ Progress through innovation and adaptation

### Part B – Short Answer

Why must students learn more than just how to use technology?

How do teachers support ethical thinking?

What are the principles of ethical stewardship?

Give one example of human–machine collaboration.

How does technology influence human evolution?

## Part C – Critical Thinking

What could happen if technology is developed without ethics?

How can education ensure a positive future with AI?

## Quiz: Chapter 14 Assessment

### Multiple Choice

Binary intelligence is the foundation of:
A. Music
B. Digital systems
C. Art
D. Sports

Digital citizenship refers to:
A. Voting
B. Online responsibility
C. Programming only
D. Hardware

Ethical stewardship includes:
A. Speed
B. Profit
C. Fairness and responsibility
D. Competition

Human–machine collaboration means:

A. Machines replace humans
B. Humans and machines work together
C. Humans stop using technology
D. Machines control humans

The future of technology depends on:

A. Machines only
B. Humans only
C. Education and ethics
D. Speed

## Short Answer

Why is balancing logic and humanity important?

How can students contribute to ethical technology use?

## ✅ Answer Key

## Worksheet Answers

### Part A

A

C

D

B

E

**Part B (Sample Answers)**
6. To understand, evaluate, and use it responsibly
7. By guiding discussions and modeling behavior
8. Fairness, transparency, inclusion, and responsibility

9. Doctors using AI for diagnosis
10. It changes how we learn, think, and solve problems

**Part C (Sample Answers)**
11. It could lead to bias, harm, and misuse
12. By teaching ethics alongside technology

### Quiz Answers

B

B

C

B

C

**Short Answer (Sample Responses)**
6. To ensure technology benefits humanity
7. By making responsible and ethical choices

# Chapter 15 – The Binary Legacy: Preserving Intelligence for Future Generations

## Overview

Every civilization leaves behind a legacy — a record of how it understood itself and the world it built.
The legacy of the 21st century will be defined not by monuments of stone or steel, but by **systems of intelligence** — codes, networks, and algorithms that shape how humanity thinks and connects.

This chapter explores how to preserve this legacy with responsibility and wisdom. It calls on educators, innovators, and future leaders to ensure that the **binary foundation of intelligence** remains a tool of empowerment, creativity, and ethical progress for generations to come.

### Learning Objectives

By the end of this chapter, students will be able to:

1. Understand the concept of the binary legacy and its impact on society.
2. Explain why knowledge preservation is essential in the digital age.
3. Describe the role of education, ethics, and archives in sustaining intelligence.
4. Reflect on personal and collective responsibility for maintaining digital integrity.
5. Envision how the next generation can expand the boundaries of intelligent systems.

## 15.1 The Meaning of Legacy in the Digital Age

Legacy once meant passing down traditions, values, and wisdom through words and culture. Today, it includes the transfer of **data, code, and digital consciousness**.

The binary legacy is the story of how humanity learned to express thought through technology — how 1s and 0s became the language of invention.
Each innovation, from the first transistor to the latest quantum processor, carries a piece of this collective memory.

Preserving it means protecting not just *machines*, but the *mindset* that created them — curiosity, ethics, and imagination.

### 15.2 Digital Preservation: The Memory of Civilization

Every document, image, and program created today contributes to the permanent digital record of humanity.
However, technology evolves quickly, and without preservation, valuable information can vanish as systems change.

To protect this legacy, societies must:

- Maintain open and accessible digital archives.
- Use standardized data formats that last across generations.
- Preserve educational resources and scientific data for global access.
- Promote policies that ensure digital history remains public and inclusive.

A well-preserved digital archive becomes a **time capsule of intelligence**, allowing future generations to learn from the past and continue its innovation.

### 15.3 Education as the Keeper of Knowledge

Schools, universities, and libraries play a vital role as the **guardians of digital wisdom**.
They do more than teach — they *curate* the legacy of human understanding.

By integrating computational literacy into curricula, education ensures that students don't just consume knowledge but **carry it forward**.

A school that teaches binary logic is not only preparing coders — it's preparing **stewards of civilization's digital memory**.

### 15.4 The Ethics of Inheritance

Inheriting intelligence comes with moral responsibility.
As technology grows more powerful, humanity must ask: *What kind of world are we leaving behind?*

The ethical preservation of intelligence means:

- Protecting digital privacy and human rights.
- Ensuring that AI serves all people, not only a few.
- Guarding against misinformation and algorithmic manipulation.
- Preserving truth in an age of synthetic content.

Just as ancient societies preserved their stories through art and scrolls, we must preserve **truth** in our digital creations.

### 15.5 Passing the Torch of Binary Wisdom

Each generation refines what it receives.
Today's students will become tomorrow's engineers, teachers, and creators — the next architects of intelligence.
They will improve on what we've built, guided by one principle: **wisdom is knowledge with purpose**.

To pass the torch means to share not only information but *understanding*.
The true legacy of the binary age lies not in machines that can think, but in **humans who continue to learn**.

## 15.6 The Global Responsibility of Intelligence

The binary legacy belongs to everyone.
Intelligence does not recognize borders, languages, or economies — it connects all people through the shared language of information.
From classrooms in small towns to research centers in major cities, each contribution adds to the world's collective mind.

Preserving this global intelligence requires cooperation across nations — open science, shared ethics, and education that transcends barriers.
The binary legacy must be built on **unity, not division**.

## 15.7 The Environmental Legacy of Computation

The growth of AI and computing also impacts the planet.
Data centers consume vast amounts of energy; electronics require rare materials.
Preserving intelligence also means **preserving the Earth that sustains it**.

Students and scientists must explore **sustainable computing** — energy-efficient chips, recycling of e-waste, and renewable power for AI infrastructure.
The most intelligent system is one that coexists with the ecosystem that gives it life.

### 15.8 The Human Legacy in the Age of AI

As AI systems become more advanced, they will carry traces of human thought — our creativity, biases, and emotions.
Each dataset, model, and design decision reflects who we are.
Therefore, our moral values will echo through the algorithms that shape the future.

By embedding compassion, fairness, and honesty into our systems, we ensure that our technological descendants inherit the best of humanity — not its flaws.

In this way, **the binary legacy becomes a mirror of our collective conscience**.

## 15.9 Looking Forward: The Continuum of Intelligence

The binary foundation is not an ending but a beginning — a bridge toward quantum, biological, and cognitive intelligence yet to come.
Just as fire, printing, and electricity changed civilization, the mastery of information will define the next century.

The question for the future is not *how smart our machines become*, but *how wise we remain*.
The evolution of intelligence will depend on humanity's ability to unite reason with empathy, code with compassion, and logic with life.

## In-Class Activities

1. **Digital Time Capsule:** Students create a file, message, or project to preserve what they believe future generations should know about the binary age.
2. **Ethical Inheritance Discussion:** Debate what responsibilities come with inheriting advanced technology.
3. **Global Legacy Map:** Identify how different countries contribute to digital preservation and innovation.
4. **Sustainability Project:** Research energy-efficient solutions for data storage and AI systems.

"The future of intelligence is not written in code alone — it is written in the choices we make."

# 📘 Chapter 15 – The Binary Legacy: Preserving Intelligence for Future Generations

## Chapter Summary

The legacy of the modern world is no longer defined by physical monuments, but by systems of intelligence—data, algorithms, and digital networks that shape how humanity thinks and communicates.

The concept of a binary legacy represents how humans have transformed simple binary code into powerful systems that store knowledge, drive innovation, and connect the world. Preserving this legacy means protecting both the technology and the values behind it.

Digital preservation plays a critical role in maintaining this legacy. As technology evolves, important information can be lost if not properly stored and maintained. Societies must create accessible archives, use long-lasting formats, and ensure knowledge remains available for future generations.

Education serves as the foundation for preserving intelligence. Schools and institutions act as guardians of knowledge, preparing students not only to understand information but to carry it forward responsibly.

Ethical responsibility is essential in maintaining this legacy. Issues such as privacy, misinformation, and fairness must be addressed to ensure that technology benefits all people.

The responsibility of preserving intelligence is global. Collaboration across countries and communities ensures that knowledge is shared and protected for everyone.

At the same time, technological growth must be balanced with environmental responsibility. Sustainable practices in computing are necessary to protect the planet while advancing innovation.

Ultimately, the binary legacy reflects humanity itself. The values embedded in today's systems will shape the future of intelligent technologies. The true measure of progress lies not in how advanced our systems become, but in how wisely and ethically we use them.

## Reflection Section

### Student Reflection

Respond to the following questions in complete sentences:

What is the binary legacy?

Why is digital preservation important?

How does education help preserve knowledge?

What ethical responsibilities come with advanced technology?

Why is global cooperation important in preserving intelligence?

How does technology impact the environment?

What kind of legacy should humanity leave for future generations?

## 📄 Worksheet: Preserving the Binary Legacy

### Part A – Vocabulary Matching

Match each term with the correct definition:

A. Binary Legacy
B. Digital Preservation
C. Ethical Responsibility
D. Global Cooperation
E. Sustainable Computing

___ Protecting and maintaining digital information

___ Working together across nations to share knowledge

___ Using technology in environmentally responsible ways

___ The record of human intelligence through digital systems

___ Using technology with fairness and integrity

### Part B – Short Answer

Why can digital information be lost over time?

How can societies preserve digital knowledge?

What role do schools play in maintaining intelligence?

Why must technology be used ethically?

How can computing become more sustainable?

## Part C – Critical Thinking

What could happen if humanity fails to preserve its digital knowledge?

How can future generations improve the binary legacy?

## 📝 Quiz: Chapter 15 Assessment

### Multiple Choice

The binary legacy refers to:
A. Physical buildings
B. Digital systems and intelligence
C. Sports records
D. Music history

Digital preservation helps:
A. Delete data
B. Store information temporarily
C. Protect knowledge for the future
D. Increase speed

Education is important because it:
A. Replaces technology
B. Preserves and transfers knowledge
C. Eliminates data
D. Reduces learning

Ethical responsibility ensures:
A. Faster machines
B. Fair and responsible use of technology

C. More storage
D. Less innovation

Sustainable computing focuses on:
A. Speed
B. Cost
C. Environmental impact
D. Design

### Short Answer

Why is global cooperation important in technology?

What role do values play in shaping the future of AI?

## ✅ Answer Key

### Worksheet Answers

**Part A**

B

D

E

A

C

**Part B (Sample Answers)**
6. Because technology changes and formats become outdated
7. By creating archives and using stable formats
8. They teach and protect knowledge
9. To prevent harm and ensure fairness
10. By using energy-efficient systems and recycling materials

**Part C (Sample Answers)**
11. Knowledge could be lost, limiting progress
12. By improving technology responsibly and ethically

## Quiz Answers

B

C

B

B

C

**Short Answer (Sample Responses)**
6. It allows shared knowledge and innovation
7. They guide how technology is created and used

# Teacher's Overview Page

## Book Title:

**The Binary Foundation of Intelligence**

## Author:

Susie Hala

## Grade Level:

Grades 6–12 (Adaptable for early college and STEM programs)

## Subject Areas:

Computer Science · Artificial Intelligence · Ethics in Technology · Digital Literacy · STEM Integration

## Purpose of This Textbook

*The Binary Foundation of Intelligence* introduces students to the logic that powers every modern technology — **binary systems**.
It connects computation, cognition, and ethics in a way that's engaging, human-centered, and cross-disciplinary.

This textbook prepares students to:

- Understand the language of logic (1s and 0s).
- Explore how intelligence evolves in both machines and nature.
- Build ethical awareness in the age of automation.
- Recognize connections between science, humanity, and innovation.

### Instructional Design

This textbook is built to align with **Next Generation Science Standards (NGSS)** and **Computer Science Teachers Association (CSTA)** digital literacy frameworks.
Each chapter includes:

- **Learning Objectives** (clear academic goals)
- **Key Concepts** (core ideas for mastery)
- **In-Class Activities** (hands-on learning and group engagement)
- **Summary Reflections** (for writing or discussion)
- **Ethical Insights** (to connect technology to human responsibility)

Teachers can use it in:

- **STEM labs and computer science classes**
- **Ethics and technology electives**
- **Cross-disciplinary lessons** linking science, math, and social studies
- **College-prep enrichment programs**

### Learning Goals

By the end of this course, students will be able to:

1. **Explain Binary Logic:** Describe how digital information is stored, processed, and communicated using 1s and 0s.
2. **Connect Logic to Life:** Relate binary reasoning to biological, cognitive, and ethical systems.
3. **Understand Artificial Intelligence:** Recognize how algorithms, data, and machine learning use binary foundations.
4. **Apply Computational Thinking:** Break down problems using logic, sequencing, and feedback loops.
5. **Develop Ethical Awareness:** Evaluate how technology impacts society, privacy, and human values.

## Suggested Chapter Groupings

| Unit | Focus Area | Chapters | Classroom Emphasis |
|---|---|---|---|
| **Unit 1 – The Language of Logic** | Foundations of binary, logic gates, and duality | 1–4 | Core concepts of binary and digital reasoning |
| **Unit 2 – Algorithms and Data** | Programming, logic structures, and decision-making | 5–9 | Application and analysis |
| **Unit 3 – Ethics and Human Context** | Moral reasoning, emotion, and machine learning | 10–12 | Interdisciplinary ethics, psychology, and AI |
| **Unit 4 – Bio-Intelligence and Living Systems** | DNA, biology, and the fusion of life and code | 13–14 | Science-based inquiry and bio-AI parallels |
| **Unit 5 – Education and Legacy** | Future thinking, sustainability, and human purpose | 15–16 | Project-based reflection and innovation |

## Assessment Ideas

- **Binary Challenges:** Students decode patterns or logic puzzles.
- **Project-Based Learning:** Design a "Binary Logic in Life" poster or AI ethics proposal.
- **Reflection Essays:** Write on dual intelligence or ethical computing.
- **Group Presentations:** Explain how binary applies to biology or art.
- **Quizzes and Tests:** Assess key vocabulary and logic comprehension.

Formative and summative assessments can be aligned with classroom goals, state standards, or AP Computer Science Principles competencies.

## Differentiation Strategies

- **For Advanced Students:** Introduce coding exercises, logic circuits, or quantum computing parallels.
- **For Beginners:** Use visual logic blocks, flowcharts, and simple analogies (e.g., light switches for 1/0).
- **For ESL Learners:** Reinforce technical vocabulary through bilingual glossaries and visual aids.
- **For Cross-Curriculum Integration:** Pair chapters with biology (DNA logic), math (Boolean algebra), and ethics (AI in society).

## Teacher Reflection Questions

1. How can binary thinking help students understand both technology and human decision-making?
2. Which ethical topics are most relevant to your classroom context?
3. How can you use binary logic as a bridge between STEM and humanities subjects?
4. What future skills — logical, emotional, or creative — should this textbook cultivate in your students?

## Final Note for Educators

*The Binary Foundation of Intelligence* empowers teachers to explore **how intelligence grows — in machines, in nature, and in ourselves**.
It encourages a holistic view of education, blending science and ethics, logic and empathy, technology and humanity.

By teaching this material, educators become not just instructors of binary code, but **mentors of future thinkers** — guiding students toward a future where intelligence is used not merely to compute, but to **care**.

# 🔤 *AI Glossary —*

## *Letter A*

### Activation Function

A key part of neural networks that decides whether a neuron should be "activated" or not. It helps the network handle complex patterns, turning raw numbers into meaningful signals during learning.

### Algorithm

A step-by-step set of instructions that tells a computer how to solve a problem or perform a task. In AI, algorithms are the "recipes" that guide machines in recognizing patterns, learning from data, and making predictions.

## Anomaly Detection

A technique used in AI to find unusual patterns or behaviors that don't fit expected data. It's commonly used in fraud detection, cybersecurity, and system monitoring to identify potential problems

### Artificial Intelligence (AI)

The science of creating machines that can perform tasks that normally require human intelligence — such as understanding language, recognizing images, solving problems, or making decisions.

### Artificial Neural Network (ANN)

A system inspired by how the human brain works, made up of layers of connected nodes (neurons). Each node processes information and passes it along, allowing the system to "learn" patterns from data.

### Artificial General Intelligence (AGI)

A theoretical form of AI that would be able to understand, learn, and perform any intellectual task a human can do. AGI doesn't exist yet, but it represents the ultimate goal of AI research — true human-like intelligence.

### Artificial Narrow Intelligence (ANI)

AI designed for a specific task, such as language translation, facial recognition, or driving a car. Most AI today is narrow — powerful at one thing, but unable to perform unrelated tasks.

### Artificial Superintelligence (ASI)

**Definition:**
Artificial Superintelligence (ASI) refers to a hypothetical form of AI that would **surpass human intelligence in all areas**, including reasoning, creativity, problem-solving, and decision-making.

**Example:**
An advanced AI that could solve complex global challenges, invent new technologies, and think far beyond human capabilities.

**Key Idea:**
ASI is **beyond human intelligence**—it represents a future level of intelligence that does not yet exist.

**Attention Mechanism**

A method used in advanced AI models (like transformers) that allows the system to focus on the most relevant parts of the input. It helps the AI "pay attention" to important words or data points when generating responses.

**Augmented Intelligence**

AI systems designed not to replace humans, but to assist them. Augmented intelligence emphasizes collaboration — combining human creativity and judgment with machine speed and accuracy.

**Autonomous System**

A system that can operate, make decisions, or take actions without direct human control. Examples include self-driving cars, delivery drones, or robotic assistants that use AI to navigate and respond to their environment.

## *Letter B*

**Backpropagation**

A key learning process used in neural networks to adjust the model's internal settings (weights) based on the errors it makes. By "propagating" mistakes backward through the network, the AI learns to improve accuracy over time.

**Batch Learning**

A method where an AI model is trained on a complete dataset all at once, rather than learning continuously. It's efficient for large, static datasets but less adaptable to changing data over time.

**Bayesian Network**

A graphical model that represents the relationships between variables and their probabilities. It helps AI systems reason under uncertainty — for example, predicting the likelihood of an event based on known factors.

**Behavioral Analytics**

AI techniques that analyze user actions, preferences, or digital patterns to predict future behavior. Often used in marketing, fraud detection, and personalized recommendations.

**Bias in AI**

When an AI system produces unfair or unbalanced results due to biased data or flawed training methods. Bias can affect hiring systems, loan approvals, and facial recognition — making fairness testing essential in AI development.

**Big Data**

Extremely large and complex datasets that traditional computers can't easily process. AI tools use big data to learn patterns, make predictions, and uncover insights across industries like healthcare, finance, and marketing.

**Binary Classification**

A machine learning task where the AI must choose between two possible outcomes — such as "spam or not spam," "yes or no," or "disease or no disease." It's one of the most basic types of AI decision-making.

**Bit**

The smallest unit of digital data, representing a value of **0** or **1**. In AI and computing, bits form the foundation for all data processing, storage, and communication. Eight bits make up one **byte**, and together they represent everything from numbers and text to images and videos.

**Bitwise Operation**

A type of mathematical operation that works directly on the binary representation (bits) of data. Bitwise operators (such as AND, OR, XOR, and NOT) are used in AI programming to control low-level logic, optimize performance, and manipulate data efficiently.

**Blockchain AI**

The combination of blockchain (a secure, distributed ledger) and artificial intelligence to improve transparency, data security, and trust in AI-driven systems.

**Bot**

Short for "robot," a bot is a program that performs automated tasks. In AI, chatbots are designed to simulate human conversation, while other bots perform repetitive digital actions like data scraping or scheduling.

**Brute Force Algorithm**

A simple but computationally expensive problem-solving method that tries every possible solution until it finds the correct one. Though inefficient, it guarantees accuracy for small problems.

## *Letter C*

## Chatbot

An AI-powered program designed to simulate human conversation through text or voice. Chatbots can answer questions, provide customer support, teach lessons, or help automate business communication using natural language processing.

## Classification

A core machine learning task where AI assigns input data into specific categories or labels — such as identifying whether an email is "spam" or "not spam." Classification models are trained using labeled examples to recognize patterns.

## Clustering

An unsupervised learning technique where AI groups similar pieces of data together based on shared characteristics. Unlike classification, clustering doesn't use labeled data — it helps uncover hidden patterns or natural groupings within datasets.

## Cognitive Computing

A branch of AI that mimics human reasoning, perception, and decision-making. Cognitive systems analyze unstructured data, recognize language and emotions, and provide intelligent suggestions — much like how the human brain interprets information.

## Computer Vision

A field of AI that enables machines to "see" and interpret visual information such as images or videos. Computer vision powers facial recognition, object detection, medical imaging, and autonomous vehicles.

## Convolutional Neural Network (CNN)

A specialized type of neural network used primarily for image and video analysis. CNNs automatically detect visual features like edges, shapes, and textures, allowing machines to recognize objects or patterns with high accuracy.

## Context Window

The range of text or tokens that a language model can "see" and use at one time to generate its response. A larger context window allows AI to remember more of the conversation, leading to better understanding and consistency.

## Corpus

A large and structured collection of texts used to train or evaluate language models. In AI, a corpus provides the examples that teach the model how words and phrases are used in real-world contexts.

### Cross-Validation

A model evaluation technique used in machine learning to test how well a model generalizes to new, unseen data. The dataset is divided into parts (folds), and the model is trained and tested multiple times to ensure accuracy and reliability.

### Cybernetics

The interdisciplinary study of communication and control in both machines and living organisms. In AI, cybernetics explores feedback loops — how systems monitor and adjust their behavior to maintain balance and achieve goals.

## *Letter D*

### Data Analytics

The process of examining, cleaning, and interpreting data to uncover useful information and patterns. In AI, data analytics helps models learn from real-world data to make predictions and informed decisions.

### Data Labeling

The act of tagging or annotating raw data with meaningful labels — such as identifying images with "cat" or "dog." Labeled data is essential for training supervised machine learning models.

### Data Mining

The practice of exploring large datasets to discover hidden patterns, trends, or relationships. AI-powered data mining helps businesses predict customer behavior, detect fraud, and improve decision-making.

### Data Science

An interdisciplinary field that combines statistics, programming, and AI to extract insights from data. Data scientists build predictive models, analyze trends, and guide data-driven decisions.

### Dataset

A structured collection of data used to train or test AI models. A dataset can include text, images, numbers, or audio. The quality and diversity of the dataset directly affect how well an AI performs.

### Decision Tree

A visual and logical model used in AI for decision-making. Each branch represents a question or condition, and each leaf represents an outcome. Decision trees are popular for their simplicity and interpretability.

### Deep Learning

A subset of machine learning that uses multi-layered neural networks to learn complex patterns. Deep learning powers technologies like voice recognition, autonomous driving, and image generation.

### Dimensionality Reduction

A mathematical process used to simplify datasets by reducing the number of variables while preserving important information. It helps AI models work faster and more efficiently, especially with large data.

### Domain Adaptation

A technique in AI that allows a model trained on one dataset to perform well on a different but related dataset. It helps AI generalize better to new environments or data types.

### Drift Detection

The process of identifying when an AI model's performance declines because real-world data has changed. Detecting data drift ensures the model stays accurate and up-to-date over time.

## *Letter E*

### E-book Recommendation Systems

AI algorithms that suggest books or reading materials based on user preferences, past selections, or browsing behavior. They enhance reading platforms by delivering personalized content to readers.

### E-Brainstorming

The use of AI tools to assist with creative idea generation. E-brainstorming platforms help individuals and teams generate fresh perspectives, organize thoughts, and develop new solutions quickly.

### E-commerce AI

AI applications that enhance online shopping experiences. These systems recommend products, predict customer behavior, optimize pricing, and improve inventory management for e-commerce platforms.

## Edge AI

Artificial Intelligence that runs directly on local devices (like smartphones, cameras, or IoT sensors) instead of relying on cloud servers. Edge AI allows faster responses, improved privacy, and reduced internet dependency.

## E-Learning AI

The use of AI in online education to personalize learning experiences, analyze student performance, and recommend tailored study materials. E-Learning AI helps teachers and learners adapt lessons to individual needs.

## E-mail Filtering AI

AI systems that automatically detect and organize incoming emails — such as sorting spam, prioritizing important messages, or detecting phishing threats — using natural language and pattern recognition.

## Embedding

A mathematical representation of data (like words, sentences, or images) in numerical form. In AI, embeddings help models understand meaning, context, and similarity — crucial for natural language processing and image recognition.

## Emotion AI

Also called Affective Computing, it's a branch of AI that recognizes, interprets, and responds to human emotions using facial expressions, voice tones, or physiological signals. It's used in customer service, education, and healthcare.

## Ensemble Learning

A technique that combines multiple AI models to produce stronger, more accurate results than a single model could achieve alone. Common ensemble methods include bagging, boosting, and stacking.

## E-Transaction Security

AI-based systems that monitor and protect online financial transactions. They detect unusual activity, verify identities, and prevent fraud in digital banking and online payment systems.

*Letter F*

## Face Recognition AI

A technology that uses machine learning and computer vision to identify or verify a person's identity based on facial features. It's used in security systems, smartphones, and law enforcement, but also raises privacy and ethical concerns.

## False Positive

A situation in which an AI system incorrectly identifies something as true when it's not. For example, a spam filter marking a legitimate email as spam. Reducing false positives improves reliability and user trust.

## False Negative

The opposite of a false positive — when an AI system fails to identify something that is true. For example, a medical AI missing signs of a disease. Balancing false positives and negatives is vital in sensitive applications.

## Feature Engineering

The process of creating, transforming, or selecting data features to improve machine learning model performance. It's a key step that combines human insight with data science techniques to optimize accuracy.

## Feature Extraction

The process of identifying the most important characteristics or variables from raw data. In AI, feature extraction helps reduce complexity and improve model accuracy by focusing on relevant information.

## Feature Map

In deep learning, particularly in convolutional neural networks (CNNs), a feature map represents visual patterns detected in an image. It helps the network recognize shapes, textures, and other important details.

## Federated Learning

A training method that allows multiple devices or organizations to collaboratively train an AI model without sharing raw data. Each participant trains the model locally, and only updates are shared — improving data privacy and security.

## Fine-Tuning

A process of retraining a pre-trained AI model on a smaller, specific dataset to make it more accurate for a particular task. Fine-tuning allows developers to customize general-purpose models for specialized applications.

### Forecasting Model

An AI system that predicts future events or trends based on historical data. Forecasting models are widely used in business, weather prediction, finance, and supply chain management.

### Fuzzy Logic

An AI reasoning system that allows for partial truths rather than strict "true or false" decisions. It's useful for handling uncertainty in real-world situations like temperature control, medical diagnosis, and robotics.

## *Letter G*

### GAN (Generative Adversarial Network)

A type of deep learning model with two neural networks — a **generator** and a **discriminator** — that work against each other. The generator creates data (like images or text), while the discriminator evaluates it for authenticity. This competition helps the model generate highly realistic outputs such as AI-generated art or photos.

### General AI (Artificial General Intelligence)

A theoretical form of AI capable of understanding, learning, and applying knowledge across multiple domains — much like a human being. Unlike narrow AI, General AI could perform any intellectual task without specific training.

### Genetic Algorithm

An optimization technique inspired by natural selection. It uses processes like mutation, crossover, and selection to evolve solutions to complex problems — often used in robotics, scheduling, and engineering design.

### Geospatial AI

The combination of artificial intelligence with geographic data (maps, satellite images, or GPS data) to analyze spatial patterns. It's used in urban planning, agriculture, environmental monitoring, and disaster management.

### Gesture Recognition

A branch of computer vision that interprets human movements — such as hand waves or facial expressions — as commands. It's commonly used in gaming, virtual reality, and touchless interfaces.

## GPT (Generative Pre-trained Transformer)

A family of large language models developed by OpenAI. GPT models are trained on vast text datasets and can generate human-like text, write essays, code, or answer questions — forming the foundation for conversational AI systems like ChatGPT.

## Gradient Descent

An optimization algorithm used in training neural networks. It minimizes the model's prediction error by gradually adjusting weights in the direction of the steepest decrease in loss — similar to rolling downhill to find the lowest point.

## Graph Neural Network (GNN)

A type of neural network designed to process data structured as graphs — with nodes (points) and edges (connections). GNNs are used in recommendation systems, molecule analysis, and social network modeling.

## Ground Truth

The real, verified data used to measure how accurate an AI model's predictions are. For example, in image recognition, the "ground truth" is the actual label of what the image shows. It's essential for evaluating model performance.

## Guided Learning

An AI-assisted educational approach that provides learners with personalized hints, explanations, and progress tracking. Guided learning uses data analytics and adaptive models to support student growth in real time.

## *Letter H*

## Hallucination (in AI)

A phenomenon where an AI model generates information that appears confident and factual but is actually incorrect or fabricated. Hallucinations occur when the model fills gaps in knowledge or misinterprets data. Managing them requires clear prompts, human review, and fact-checking.

## Heuristic Algorithm

A problem-solving method that uses experience-based shortcuts to find good — though not always perfect — solutions quickly. Heuristic algorithms are common in AI search systems, decision-making, and optimization problems where speed matters more than precision.

## Hidden Layer

A layer of neurons in a neural network that processes inputs between the input and output layers. Hidden layers extract complex patterns from data and enable the network to make accurate predictions or classifications.

## Human-Computer Interaction (HCI)

The study and design of how people interact with computers and AI systems. HCI focuses on usability, accessibility, and user experience, ensuring that AI tools feel intuitive and supportive to human users.

## Human-in-the-Loop (HITL)

An AI training or decision process that includes active human participation. Humans review, correct, or guide the AI's outputs to ensure quality, fairness, and ethical alignment — especially important in high-stakes fields like healthcare and law.

## Hybrid AI

An approach that combines two or more types of AI — such as symbolic reasoning (rule-based) and machine learning (data-driven) — to create systems that are both interpretable and powerful. Hybrid AI blends logic with learning for smarter, explainable outcomes.

## Hyperparameter

A setting or configuration that controls how an AI model learns, such as learning rate, batch size, or number of layers. Unlike model parameters learned automatically, hyperparameters are manually chosen to optimize training performance.

## Hyperparameter Tuning

The process of finding the best hyperparameter values for a model. It's often done through techniques like grid search, random search, or Bayesian optimization to achieve the highest accuracy possible.

## Hypothesis Testing (in AI)

A statistical method for evaluating whether a pattern or relationship found by an AI model is real or due to chance. It helps ensure that model results are meaningful and scientifically valid.

## Hybrid Cloud AI

A cloud computing setup that uses both private and public cloud environments to train and deploy AI models. It offers flexibility, scalability, and stronger control over sensitive data.

## *Letter I*

## Image Recognition

A branch of computer vision that enables AI systems to identify and categorize objects, people, or scenes in images. It's used in facial recognition, medical imaging, self-driving cars, and quality control in manufacturing.

## Imbalanced Data

A situation where some classes or categories in a dataset have many more examples than others. Imbalanced data can cause AI models to favor the majority class, leading to inaccurate predictions. Balancing data improves fairness and performance.

## Inference Engine

A component of an AI system that applies logical rules or learned patterns to reach conclusions or make decisions. Inference engines are core parts of expert systems and modern language models that generate intelligent responses.

## Information Retrieval

The process of searching, extracting, and ranking relevant data from large datasets or documents. AI-powered search engines and chatbots use information retrieval to deliver accurate, context-aware answers.

## Input Data

The raw information — such as text, numbers, or images — that an AI system receives before processing. The quality of input data directly affects the reliability of the model's output.

## Intelligent Agent

A software entity capable of perceiving its environment, making decisions, and taking actions to achieve specific goals. Examples include virtual assistants, recommendation systems, and AI tutors.

## Interpretable AI

An approach to AI design focused on making model decisions understandable to humans. Interpretable AI helps users trust and verify machine-generated outcomes, especially in healthcare, finance, and law.

### Iterative Learning

A process where an AI model improves through repeated cycles of training, testing, and refining. Each iteration helps the system learn from its mistakes and perform better over time.

### IoT (Internet of Things) AI

AI integrated with smart devices connected through the internet. IoT AI enables objects like thermostats, watches, and appliances to learn user preferences, predict behavior, and operate autonomously.

### Isolation Forest

An anomaly detection algorithm that identifies unusual data points by isolating them from the rest of the dataset. It's efficient for spotting fraud, system errors, or rare events.

## *Letter J*

### Jaccard Index

A statistical measure used to compare the similarity between two sets. In AI and machine learning, it's often used to evaluate clustering or text similarity by dividing the intersection of sets by their union.

### Java (in AI)

A programming language often used in AI applications due to its portability, scalability, and strong memory management. Java supports frameworks like Deeplearning4j and Weka, which help in building AI systems for enterprise and academic use.

### Joint Embedding

A technique where two or more types of data — such as text and images — are represented in a shared vector space. Joint embeddings allow AI to link visual and textual understanding, as seen in models like CLIP that connect words to images.

### JSON (JavaScript Object Notation)

A lightweight data format used for storing and exchanging structured information between systems. AI tools often use JSON to send model inputs, outputs, and metadata in a clear and readable way.

## Jump Learning

An advanced learning concept where AI "jumps ahead" in reasoning or skill based on prior knowledge, instead of learning everything from scratch. It mimics human intuition and helps reduce training time in complex models.

## *Letter K*

## Keyword Extraction

A natural language processing technique used to identify the most relevant words or phrases within text. It helps summarize content, improve search results, and enhance AI-driven document classification.

## Kernel Function

A mathematical function used in Support Vector Machines (SVMs) to transform data into a higher-dimensional space. This allows the AI to separate complex, non-linear data more effectively. Common kernel types include linear, polynomial, and radial basis functions.

## Kinetic AI

A branch of AI that studies movement and motion analysis. It's used in robotics, sports technology, and autonomous vehicles to understand physical actions and predict motion patterns.

## K-Means Clustering

An unsupervised learning algorithm that divides data into *K* distinct groups based on similarity. It assigns each data point to the cluster with the nearest mean value. K-Means is widely used in customer segmentation, image compression, and pattern recognition.

## K-Nearest Neighbor (KNN)

A simple yet powerful machine learning algorithm that classifies data based on its proximity to *K* neighboring points. It assumes similar data points are likely to have similar outcomes. KNN is used in recommendation systems and predictive analytics.

## Knowledge Base

A structured collection of information that an AI system uses to answer questions, make decisions, or provide explanations. Knowledge bases power systems like virtual assistants, expert systems, and search engines.

## Knowledge Distillation

A process where a smaller AI model learns from a larger, more complex model. This helps create faster, lightweight systems that maintain strong performance while using fewer computational resources.

### Knowledge Graph

A network that connects information through relationships and context, enabling AI systems to understand how data points relate. Knowledge graphs are used by companies like Google to enhance search accuracy and context understanding.

### Knowledge Representation

The method of encoding information in a way that AI systems can understand and use for reasoning. It includes logic rules, ontologies, and semantic networks, allowing AI to store and apply knowledge intelligently.

### Knowledge Transfer

A technique in AI and machine learning where knowledge gained from one task or dataset is reused in another. This reduces the need for retraining from scratch and speeds up learning in related areas.

## *Letter L*

### Label Encoding

A data preprocessing technique used to convert categorical (text-based) values into numerical form so that AI models can process them. For example, “Red,” “Green,” and “Blue” may be encoded as 0, 1, and 2.

### Labeled Data

Data that has been tagged with correct answers or classifications, such as images labeled “cat” or “dog.” Labeled data is essential for training supervised learning models that rely on known outcomes.

### Language Model

An AI system trained to understand, predict, and generate human language. It learns from massive text datasets to perform tasks like translation, summarization, and conversation — examples include GPT and Gemini.

### Latent Space

A hidden mathematical representation within an AI model that captures complex relationships between data points. In generative AI, latent space allows systems to create new images, sounds, or text by exploring abstract features.

## Lazy Learning

A machine learning approach where the model delays generalization until a prediction is requested. Algorithms like K-Nearest Neighbor (KNN) store data and only compute results when needed, trading speed for flexibility.

## Linear Regression

A basic yet powerful algorithm that predicts outcomes by finding the best-fitting straight line through data points. It's commonly used in finance, sales forecasting, and scientific analysis.

## Logistic Regression

A classification algorithm used to predict categorical outcomes (like "yes" or "no") instead of continuous values. Despite its name, it's mainly used for probability-based classification rather than regression.

## Long Short-Term Memory (LSTM)

A special type of recurrent neural network (RNN) designed to remember long-term dependencies in sequential data, such as language or time series. LSTMs are used in speech recognition, translation, and text generation.

## Loss Function

A mathematical formula that measures how far an AI model's predictions are from the correct answers. The goal of training is to minimize this loss, helping the model improve accuracy over time.

## Low-Code AI

AI tools that allow users to build and deploy machine learning models with minimal coding. They simplify development for non-programmers, empowering educators, entrepreneurs, and small businesses to use AI effectively.

## *Letter M*

## Machine Learning (ML)

A core branch of AI where systems learn from data to make predictions or decisions without being explicitly programmed. It powers tools like recommendation engines, fraud detection, and voice recognition.

### Meta Learning

Also called "learning to learn," this technique trains AI systems to adapt quickly to new tasks by using prior knowledge. Meta learning helps models generalize better with less training data.

### Model Training

The process of teaching an AI model to recognize patterns by feeding it large amounts of data. During training, the model adjusts its internal parameters to minimize errors and improve predictions.

### Model Evaluation

A step where an AI model's accuracy and performance are tested using data it hasn't seen before. Evaluation ensures that the model works well in real-world scenarios.

### Model Overfitting

A situation where an AI model learns too closely from training data, capturing noise instead of patterns. Overfitting causes poor performance on new data. Regularization and validation help prevent it.

### Model Optimization

The process of improving an AI model's efficiency, accuracy, and speed by fine-tuning its architecture, parameters, or algorithms.

### Monte Carlo Simulation

A statistical method that uses random sampling to estimate outcomes in uncertain situations. AI systems use it for risk analysis, forecasting, and game simulations.

### Multimodal AI

An AI system capable of understanding and processing information from multiple sources — such as text, images, sound, and video — to produce richer and more accurate results.

### Multi-Agent System

A group of AI agents that work together or compete to achieve individual or shared goals. They are used in simulations, logistics, and automated trading systems.

## Mutation Algorithm

A process inspired by genetics where random changes are introduced to improve AI models' performance during optimization or evolutionary learning.

*Letter N*

## Naive Bayes Algorithm

A simple but powerful machine learning algorithm based on probability. It's widely used in spam detection, text classification, and sentiment analysis.

## Named Entity Recognition (NER)

An NLP technique that identifies and classifies specific names or entities (like people, places, or organizations) within text.

## Natural Language Processing (NLP)

A field of AI that enables machines to understand and respond to human language. It powers chatbots, translation tools, and voice assistants.

## Natural Language Understanding (NLU)

A subset of NLP that focuses on interpreting meaning, emotion, and intent behind text or speech. It allows AI to understand context and tone.

## Neural Architecture Search (NAS)

An automated method for designing neural networks. AI systems use NAS to discover optimal model structures without human intervention.

## Neural Network

A computational model inspired by the human brain, made of layers of nodes (neurons) that process data. Neural networks are the foundation of deep learning systems.

## Neuron

## Definition

In artificial intelligence and machine learning, a **neuron** (also called a *node* or *unit*) is the fundamental building block of an artificial neural network. It is modeled loosely after the biological neuron found in the human brain. Each artificial neuron receives input signals, processes them mathematically, and produces an output that is passed to other neurons in the network. Neurons collectively enable a model to learn, recognize patterns, and make predictions by adjusting their internal parameters during training.

## Key Components

1. **Inputs:**
   The data values or signals that are fed into the neuron — similar to how biological neurons receive signals through dendrites.
2. **Weights:**
   Numerical values assigned to each input that determine their importance. The neuron learns by adjusting these weights during training.
3. **Summation Function:**
   A mathematical operation that combines the weighted inputs into a single value, often using a simple addition or dot product.
4. **Activation Function:**
   A non-linear function (such as ReLU, Sigmoid, or Tanh) applied to the summation result to decide whether the neuron should "fire" or remain inactive.
5. **Output:**
   The final result produced by the neuron after applying the activation function, which becomes the input for other neurons in subsequent layers.

## Benefits

- Allows complex relationships to be learned from raw data.
- Enables non-linear transformations for deep learning tasks.
- Builds scalable and flexible structures that can adapt to various problems.
- Mimics the way biological neurons communicate and adapt over time.

## Disadvantages

- Individual neurons are simple and limited in function; meaningful behavior arises only when many are connected.
- May become computationally expensive as the network grows deeper.
- Can overfit data if not properly regularized.

## Examples

- A neuron in a **convolutional neural network (CNN)** may detect visual patterns such as edges or textures.
- In a **language model**, neurons may learn linguistic features like word order or meaning.
- In **speech recognition**, neurons can capture tone, pitch, and pronunciation patterns.

## Applications

- **Image Recognition:** Detecting and classifying objects, faces, or handwriting.
- **Natural Language Processing:** Understanding text, context, and sentiment.

- **Financial Forecasting:** Identifying market trends from numerical data.
- **Medical Diagnosis:** Interpreting scans, patterns, and patient records.

### Industries Using Neuron-Based Models

- **Healthcare:** Predictive diagnosis and imaging analysis.
- **Education:** Adaptive learning and student performance prediction.
- **Technology:** AI assistants, chatbots, and autonomous systems.
- **Transportation:** Self-driving cars and sensor-based navigation.

### Challenges

- Understanding neuron-level decision-making remains difficult, leading to "black box" behavior.
- Requires large amounts of data and computational resources to train effectively.
- Sensitive to data quality — poor input can lead to unreliable neuron activation.

### Summary

A **neuron** is the core unit of intelligence within an artificial neural network. By processing inputs through weighted connections and activation functions, neurons enable machines to simulate learning and reasoning similar to the human brain. Though individually simple, millions of interconnected neurons together form the foundation of modern AI systems, powering technologies from image recognition to natural language understanding.

### Node

A basic unit in a neural network that receives input, processes it, and sends it forward. Nodes collectively form layers that learn from data.

### Noise (in Data)

Random or irrelevant information in a dataset that can confuse AI models. Removing noise improves accuracy and training efficiency.

### Nonlinear Activation

A mathematical function in neural networks that allows models to learn complex, non-straight-line relationships. Common examples include ReLU, Sigmoid, and Tanh functions.

### Normalization

A data preprocessing step that scales values into a consistent range to help AI models learn faster and perform better.

*Letter O*

## Object Detection

A computer vision technique that identifies and locates objects within an image or video. It's used in autonomous driving, security systems, and retail analytics.

## Objective Function

A mathematical formula that defines the goal of an AI model — usually to minimize error or maximize accuracy. It guides the learning process during training.

## One-Hot Encoding

A method used to represent categorical data numerically by converting each value into a binary vector. It helps AI systems interpret text-based or labeled inputs.

## Ontology (in AI)

A structured framework that defines relationships between concepts within a domain. Ontologies help AI organize knowledge logically and reason effectively.

## Optimization Algorithm

A method used to adjust model parameters during training to minimize loss and improve accuracy. Examples include Gradient Descent and Adam Optimizer.

## Optimization Problem

A task in AI where the goal is to find the best possible solution — like minimizing cost or maximizing performance — under certain constraints.

## Outlier Detection

The process of identifying data points that deviate significantly from the rest. Outlier detection helps find errors, fraud, or anomalies.

## Overfitting

When an AI model performs well on training data but fails to generalize to new data because it memorized instead of learning patterns.

## Overparameterization

A condition where an AI model has more parameters than necessary, often leading to overfitting and inefficiency.

### Oversampling

A data balancing technique that increases the number of examples in an underrepresented class to prevent bias during training.

## *Letter P*

### Parameter

A value inside an AI model that determines how it makes predictions. Parameters are adjusted during training to minimize error and improve accuracy.

### Pattern Recognition

The ability of AI systems to identify regularities, trends, or structures in data. It's used in handwriting recognition, medical imaging, and speech analysis.

### Perceptron

One of the earliest types of artificial neural networks. It consists of a single layer of nodes and is used for simple classification tasks.

### Predictive Analytics

A branch of AI that uses statistical models and machine learning to forecast future events based on historical data.

### Preprocessing

The preparation of data before feeding it into an AI model — including cleaning, scaling, and encoding — to ensure accuracy and consistency.

### Probability Distribution

A statistical function that describes the likelihood of different outcomes in a dataset. It's essential for probabilistic reasoning and uncertainty modeling in AI.

### Prompt Engineering

The practice of crafting precise and intentional instructions (prompts) to guide AI models in generating accurate and creative responses. It's the foundation of effective human-AI communication.

### Probability Distribution

A statistical function that describes the likelihood of different outcomes in a dataset. It's essential for probabilistic reasoning and uncertainty modeling in AI.

### Python (Programming Language)

A versatile, easy-to-learn programming language widely used in AI and data science. Python supports major libraries like TensorFlow, PyTorch, and Scikit-learn, making it the most popular tool for AI development.

### Pre-trained Model

An AI model that has already been trained on large datasets and can be fine-tuned for specific tasks. Using pre-trained models saves time and computing resources.

### Precision and Recall

Two key metrics for evaluating AI performance. **Precision** measures accuracy among positive predictions, while **Recall** measures how many actual positives were correctly identified.

## *Letter Q*

### Q-Learning

A type of reinforcement learning algorithm where an agent learns to make decisions by receiving rewards or penalties. It helps AI systems determine the best action to take in each situation, such as in robotics or game AI.

### Quantization

A technique that reduces the precision of numbers in AI models to make them smaller and faster without major loss in accuracy. It's commonly used to optimize models for mobile or edge devices.

### Quantum AI

The integration of quantum computing and artificial intelligence. Quantum AI uses the principles of quantum mechanics to process complex computations faster than traditional computers.

### Query

A request for information sent to a database or AI system. In natural language models, queries are user inputs — the "questions" that trigger responses.

### Queue Learning

A data-handling strategy where information is processed sequentially, one at a time. It's useful for managing AI tasks or workflows that must follow a specific order.

## *Letter R*

### Random Forest

An ensemble learning algorithm that combines multiple decision trees to improve accuracy and reduce overfitting. It's used for both classification and regression tasks.

### Recurrent Neural Network (RNN)

A neural network designed for sequential data such as text, speech, or time series. RNNs have loops that allow them to "remember" previous inputs, making them ideal for language and forecasting tasks.

### Regression Analysis

A statistical method that models the relationship between variables to predict future outcomes. Linear and logistic regression are common types.

### Reinforcement Learning

An AI learning method where an agent interacts with its environment and learns from feedback through rewards and penalties. It's used in robotics, gaming, and automation.

### Representation Learning

A machine learning technique where the system automatically discovers useful features from raw data instead of relying on manual feature selection.

### Retrieval-Augmented Generation (RAG)

A model that combines information retrieval and text generation. It fetches relevant data before producing a response, improving accuracy in AI question-answering systems.

### Reward Function

A key component of reinforcement learning that defines how much "reward" an AI receives for taking certain actions. It guides the system toward desirable outcomes.

### Robotic Process Automation (RPA)

Technology that uses AI to automate repetitive digital tasks, such as data entry or document processing, to increase efficiency and reduce human error.

### Rule-Based System

An AI system that makes decisions based on predefined rules or conditions, often used in expert systems or early AI programs.

### Runtime Environment

The system or platform where AI models are executed. It includes the software, hardware, and configurations required for the model to function properly.

## *Letter S*

### Scalability

The ability of an AI system to handle increasing amounts of data or users without losing performance. Scalable AI models grow with business and computational needs.

### Semi-Supervised Learning

A learning method that combines a small amount of labeled data with a large amount of unlabeled data to improve model performance.

### Sentiment Analysis

A natural language processing technique that identifies emotions or opinions in text, such as positive, negative, or neutral sentiment. It's used in marketing and social media monitoring.

### Signal Processing

The analysis and transformation of data signals (like sound, images, or time series) into meaningful patterns. AI uses it in speech recognition and medical imaging.

### Speech Recognition

A technology that converts spoken language into text using AI models trained on voice data. It powers virtual assistants and transcription tools.

### Supervised Learning

A machine learning technique where a model is trained using labeled data, meaning both inputs and correct outputs are provided during training.

### Swarm Intelligence

An AI concept inspired by nature, where simple agents (like ants or bees) collectively solve complex problems. It's used in optimization, robotics, and network routing.

### 8. Symbolic AI

An early AI approach based on logical rules and symbolic representations. Unlike modern machine learning, it uses explicit reasoning rather than pattern recognition.

### Synthetic Data

Artificially generated data created to train AI models when real data is limited or sensitive. It helps improve privacy and expand datasets.

## *Letter T*

### Target Variable

The specific outcome or label that an AI model aims to predict, such as "spam or not spam" or "disease present or absent."

### TensorFlow

An open-source AI framework developed by Google that helps developers build and train machine learning models efficiently.

### Token Limit

The maximum number of tokens (words or symbols) an AI model can process in a single interaction or context window

### Tokenization

The process of breaking text into smaller units, or "tokens," such as words or subwords, to help AI models understand language.

### Transfer Learning

A technique where a model pre-trained on one task is adapted to a different but related task, saving time and improving performance.

## Transformer Architecture

### Definition

Transformer Architecture is a deep learning model framework introduced in 2017 by Vaswani et al. in the paper *"Attention Is All You Need."* It revolutionized natural language processing (NLP) by replacing traditional sequence models like RNNs (Recurrent Neural Networks) and LSTMs (Long Short-Term Memory networks) with a mechanism known as **self-attention**, allowing the model to understand relationships between words or elements in a sequence — regardless of their distance from each other.

### Key Components

1. **Encoder-Decoder Structure:**
   The architecture typically has two main parts — the encoder, which reads and represents the input data, and the decoder, which generates the output based on the encoded information.
2. **Self-Attention Mechanism:**
   This mechanism lets the model weigh the importance of each word relative to others in a sentence, improving context understanding.
3. **Multi-Head Attention:**
   Multiple attention layers run in parallel, enabling the model to capture different kinds of relationships or meanings simultaneously.
4. **Positional Encoding:**
   Since Transformers do not process data sequentially, positional encodings are added to indicate the order of elements in a sentence.
5. **Feed-Forward Networks:**
   After attention layers, fully connected neural networks help the model transform and refine the contextual representations.
6. **Residual Connections and Layer Normalization:**
   These ensure stability during training and help the model learn complex patterns efficiently.

### Benefits

- Handles long-range dependencies more effectively than RNNs.
- Enables parallel processing, making it faster to train on large datasets.
- Produces state-of-the-art performance across multiple tasks, such as translation, summarization, and text generation.
- Scales well to massive datasets and larger models like GPT, BERT, and T5.

### Disadvantages

- Extremely data- and computation-intensive, requiring powerful hardware.
- Can be prone to bias if trained on unbalanced or low-quality datasets.
- Difficult to interpret because of its complexity and multi-layered attention mechanisms.

## Applications

- **Natural Language Processing (NLP):** Machine translation, text summarization, sentiment analysis, and chatbots.
- **Vision Transformers (ViTs):** Image recognition and object detection in computer vision.
- **Speech Recognition:** Improved context-aware transcription and voice command systems.
- **Multimodal AI:** Combining text, images, and audio for generative and reasoning tasks.

## Industries Using Transformer Models

- **Technology:** Google, OpenAI, Meta, and Microsoft for large-scale AI systems.
- **Education:** Intelligent tutoring systems and adaptive learning platforms.
- **Healthcare:** Medical data analysis and clinical text interpretation.
- **Entertainment:** Script generation, gaming narratives, and subtitles translation.

## Challenges

- High energy consumption during training and inference.
- Potential for producing inaccurate or biased outputs if the model is not fine-tuned properly.
- Data privacy concerns in large-scale, open-domain training environments.

## Summary

The Transformer Architecture represents a fundamental shift in how machines understand and generate language. Its self-attention mechanism allows models to process context more globally and flexibly than ever before. Transformers laid the foundation for today's most advanced AI systems — including ChatGPT, BERT, and countless others — marking a turning point in modern artificial intelligence.

## Transformer Model

A neural network architecture that processes sequences of data (like text) more efficiently than traditional RNNs. It's the foundation of modern large language models such as GPT.

## Training Data

The dataset used to teach an AI model. Training data must be diverse and accurate to ensure reliable model performance.

## Turing Test

A test developed by Alan Turing to evaluate whether a machine can exhibit intelligent behavior indistinguishable from that of a human.

## Time Series Analysis

The use of AI to analyze sequential data over time, such as stock prices or weather patterns, to forecast future trends.

## Tunable Parameters

Adjustable values in an AI model that influence its performance, such as learning rate or layer depth. Tuning helps achieve optimal results.

# *Letter U*

## Unlabeled Data

Data that has not been tagged with specific outputs or categories. It's commonly used in unsupervised and semi-supervised learning.

## Unstructured Data

Information that doesn't follow a predefined format, such as text, images, videos, or social media posts. AI tools process it using natural language and computer vision techniques.

## Unsupervised Learning

A machine learning method where models learn from unlabeled data to find patterns, groupings, or hidden structures without specific instructions.

## Uncertainty Modeling

The process of estimating how confident an AI model is about its predictions. It helps improve trust and decision-making in sensitive domains like medicine or finance.

## Update Rate

The frequency at which an AI model adjusts its parameters during training. A balanced update rate ensures faster learning without instability.

## User Intent Recognition

An NLP technique that determines the purpose or goal behind a user's input — essential in chatbots and voice assistants for accurate responses.

### Utility Function

A mathematical function used in AI to evaluate the desirability or usefulness of different actions. It's often used in decision-making and reinforcement learning.

## *Letter V*

### Validation Set

A portion of the dataset used during model training to fine-tune parameters and prevent overfitting. It helps evaluate how well the model generalizes to unseen data before final testing.

### Value Function (in Reinforcement Learning)

A function that estimates the expected reward an AI agent will receive from a certain state or action. It guides decision-making by helping the agent choose actions that maximize long-term rewards.

### Variance (in AI Models)

A measure of how much a model's predictions change when trained on different data samples. High variance means the model is sensitive to noise and may overfit.

### Vectorization

The process of converting data (like text or images) into numerical vectors that AI models can process. It's essential in natural language processing and computer vision.

### Verification (in AI Systems)

The process of checking whether an AI model or system performs its intended function correctly and safely before deployment.

### Virtual Agent

An AI-driven digital assistant designed to interact with humans through voice or text. Examples include chatbots, customer service bots, and AI tutors.

### Virtual Reality (VR)

A computer-generated environment that simulates real or imagined worlds. When combined with AI, VR can create adaptive, interactive learning and training experiences.

### Vision Transformer (ViT)

A deep learning model that applies transformer architecture — originally used for language — to image recognition tasks. It divides images into patches and processes them like words in a sentence.

### Voice Recognition

AI technology that identifies and processes human voices to perform tasks, such as unlocking devices or activating voice assistants.

### Volumetric Data

Three-dimensional data used in fields like medical imaging, 3D modeling, and geoscience. AI uses it to analyze structures and visualize depth in complex environments.

## *Letter W*

### Weak AI (Narrow AI)

AI designed to perform a specific task, such as translation or image classification. It does not possess general reasoning or consciousness like humans.

### Weight (in Neural Networks)

A numerical value that determines the strength of a connection between neurons. During training, weights are adjusted to reduce prediction errors.

### Weighted Average

A statistical method that assigns different importance (weights) to data points when calculating an average. AI uses it to prioritize more relevant or reliable data.

### Whisper Model

An AI model developed by OpenAI for automatic speech recognition (ASR). It transcribes and translates spoken language into text across multiple languages with high accuracy.

### Wi-Fi Sensing AI

Technology that uses Wi-Fi signals to detect movement or monitor environments without cameras. It's used in security, healthcare, and smart homes.

### Windowing Function

A mathematical function used in signal processing and speech recognition to divide data into smaller segments for analysis.

### Word Embedding

A representation of words in numerical form where similar words are placed close together in vector space. It helps AI models understand semantic relationships in text.

### Word2Vec

A popular algorithm for creating word embeddings by learning relationships between words based on their context within large text datasets.

### Workflow Automation

The use of AI to streamline repetitive business tasks such as approvals, document sorting, and scheduling. It increases productivity and reduces human error.

### World Model (in AI)

A conceptual model an AI system builds to understand and predict the behavior of the environment around it — especially in robotics and autonomous systems.

## *Letter X*

### XAI (Explainable AI)

A field of AI focused on making models transparent and understandable to humans. It explains how decisions are made, which builds trust and accountability.

### XGBoost (Extreme Gradient Boosting)

A powerful machine learning algorithm known for its high performance in classification and regression tasks. It's widely used in data science competitions and predictive modeling.

### XML (Extensible Markup Language)

A structured text format used to store and share data between systems. AI applications use XML for configuration, annotations, and data exchange.

### XOR Problem

A classic logic problem that early neural networks struggled to solve. It led to the development of multilayer neural networks capable of learning non-linear relationships.

*Letter Y*

**YAML (YAML Ain't Markup Language)**

A human-readable data format often used for AI configuration files, datasets, and machine learning pipelines. It's easy to read and integrates well with Python-based AI frameworks.

**Yield Prediction AI**

AI systems that forecast agricultural output based on environmental data, crop type, and soil conditions. It supports smarter farming and food sustainability.

**Yearly Seasonality Model**

A model used in time-series forecasting that accounts for repeating annual patterns — such as sales spikes or weather cycles — to improve accuracy.

**YOLO (You Only Look Once)**

A fast real-time object detection algorithm that identifies and locates multiple objects in a single pass. It's widely used in autonomous vehicles and security systems.

*Letter Z*

**Zero-Shot Learning**

An AI method that allows a model to recognize new objects or tasks it has never seen before by using prior knowledge or contextual reasoning.

**Zettabyte**

A unit of digital storage equal to one sextillion ($10^{21}$) bytes. AI models and data centers increasingly operate with zettabyte-scale data due to massive global information growth.

**Z-Score Normalization**

A statistical technique used to standardize data by subtracting the mean and dividing by the standard deviation. It helps machine learning models handle data with different scales.

**Zone-Based AI**

A decision-making approach where an AI system divides its environment into regions or "zones" for localized reasoning and optimized control — used in robotics and traffic management.

**NOTES:**

**NOTES:**

www.ingramcontent.com/pod-product-compliance
Lightning Source LLC
Chambersburg PA
CBHW041754220926
51847CB00014B/201
* 9 7 8 1 9 7 2 9 2 5 0 0 3 *